# Enhancing the Polulu 3pi with RobotBASIC

## John Blankenship
## Samuel Mishal

The source code for the program listings in this book (and much
more) is available to the readers at:

http://www.RobotBASIC.com

# Contents at a Glance

# Table Of Contents

**Chapter 4: The Primary 3pi Software  29**

# Preface

RobotBASIC, a free Robot Control Language, has an integrated robot simulator with numerous sensors, making it easy to develop complex robot algorithms and behaviors. This book explains how to modify and expand the standard Pololu 3pi robot so that it has many of the capabilities of RobotBASIC's simulated robot. It also explores several example projects utilizing the modified robot in order to demonstrate how to exploit the new features.

If you have not programmed in RobotBASIC before, please visit our web page
**www.RobotBASIC.com**
to download your free copy of RobotBASIC as well as a short PDF Tutorial on using the language. RobotBASIC comes with nearly 300 pages of documentation, but if you are totally new to programming you might consider one of our introductory texts such as *RobotBASIC Projects for Beginners* or *Robots in the Classroom*, both of which are discussed and available on our web page. If you need more detailed information on developing algorithms for robotic behaviors, consider our advanced book, *Robot Programmer's Bonanza*.

The project described in this book is not suitable for beginners. The construction of the robot described here, assumes you are familiar with soldering techniques and can read and interpret schematics. If you do not possess these skills you can easily DESTROY the 3pi robot. It would be nice if Pololu could offer either a kit or fully assembled version of the robot described in this book and we will certainly help them pursue such an endeavor should they feel the demand warrants it. If you would be willing to purchase such a product, please let them know at **www.Pololu.com**.

Finally, we realize this is a daunting project that is not suitable for many non-technical hobbyists. For that reason, we are constantly working on ways to make it easier for everyone to use RobotBASIC with real-world robots. Visit our web page for the latest news and innovations.

# Chapter 1

# Overview

**A**lthough the RobotBASIC simulated robot provides a very powerful and rewarding programming experience, for many hobbyists, a real physical robot is the ultimate goal. Unfortunately, most robot kits provide only a few types of sensors at best, and if you've programmed the RobotBASIC simulator, you know you cannot give your robot intelligent behaviors without an appropriate number and variety of sensors. If you have not programmed with RobotBASIC before, it is suggested that you read the PDF Tutorial and other information available at **www.RobotBASIC.com**.

The sensors (types, number of, and locations of) provided on the RobotBASIC simulator were not chosen lightly. They provide everything necessary to develop a wide variety of robotic behaviors. It follows then, that an ideal hobby robot would have the same sensors as the RobotBASIC simulation and offer transparent compatibility with the RobotBASIC language.

Building such a robot can be a daunting task, even with today's technology. Initially, we considered building a robot totally from scratch, but when we found the Pololu

3pi, we knew we had a nearly perfect platform for a powerful educational or hobbyist-oriented robot.

The standard Pololu 3pi (see Figure 1.1) comes assembled, with five line sensors and the ability to read the battery voltage. Even though that does not sound like a lot, the line sensors allow the implementation of algorithms to follow lines and negotiate line mazes. For more information on the 3pi, visit **http://www.pololu.com.**

**Figure 1.1**: The Pololu 3pi is a small, yet ideal, starting platform for creating our robot.

Notice that the 3pi is round, just like the RobotBASIC simulation. Pololu provides software that offers a *slave* mode that makes it easy to control the 3pi from an external computer or microcontroller. While such a mode would make some implementation tasks easy (and can certainly be accomplished using the RobotBASIC serial commands), it does not provide the power nor the flexibility truly required for this project.

## 1.1 RobotBASIC's Communication Protocol

RobotBASIC provides a unique built-in serial protocol for wireless adapters (such as Bluetooth) that allows programs developed with the simulator to immediately control a real robot. Let's see how this works.

Normally, when you want to tell the simulated robot to move, you need a command like `rForward 20` which tells the simulated robot to move forward on the screen twenty pixels (negative numbers move backward). A command such as `rTurn 15` tells the robot to turn 15° to the right (negative numbers turn left).

If you want these same commands to control an external real robot, you simply have to issue the command `rCommPort N` (at the beginning of your program). This command indicates that a Bluetooth adapter (or other communication device) has been connected and is using serial port **N** (typically a Virtual Serial Port). Just replace **N** with the actual number of the port being used - more on this later.

Once the `rCommPort` command has been issued, all robot commands (including `rForward` and `rTurn`) no longer control the simulation. Instead, they automatically send two bytes to the specified serial port. The first of these bytes is an operation code that identifies the command. The table in Figure 1.2 shows the code used for each of the simulator commands we will be implementing.

| Command | Op-code |
|---|---|
| rLocate | 3 |
| rForward | 6 |
| (backward) | 7 |
| rTurn (right) | 12 |
| (left) | 13 |
| rCompass | 24 |
| rBeacon | 96 |
| rRange (right) | 192 |
| (left) | 193 |
| rSpeed | 36 |
| rChargeLevel | 108 |

**Figure 1.2**: These are the RobotBASIC commands (and their op-codes) that will be implemented on the 3pi.

The second byte sent to the robot is either zero (if not needed) or a parameter related to the command. For example, when the command rForward 20 is issued, the PC will send out a 6 followed by a 20. The 6, of course, indicates the FORWARD operation is being requested and the 20 specifies how far forward to move.

In addition to the above commands, RobotBASIC provides an rCommand function that allows the RobotBASIC programmer to implement custom commands. In a later chapter, we will see how this command can provide a number of unique capabilities.

In addition to receiving commands from the PC, the external robot has to return sensory data to RobotBASIC. The robot returns five bytes of sensory data *every* time it receives a command. Three pieces of data (information from the bumpers, the infrared object detection sensors, and the line sensors) are very time sensitive and are nearly always returned in the first three of these five bytes (in the order listed above).

The remaining two bytes are usually zero because they are typically not needed. When the commands rCompass, rBeacon, and rRange are executed though, these two bytes are used to return the requested data.

When RobotBASIC receives these five bytes, it automatically extracts the individual pieces of information and uses them appropriately. For example, the rBumper() function normally provides the status of the simulated robot's bumpers. After an rCommPort command has been issued though, rBumper will return the status of the real robot's bumpers because RobotBASIC will automatically use the last data received from the external robot.

## 1.2 Capabilities to be Implemented

When fully implemented, this provides a system where the external robot has nearly all of the capabilities of the simulation. These capabilities are summarized below.

- 4 Bumper sensors (front, rear, left, and right)
- 5 IR perimeter proximity sensors (all equally spaced on the front half of the robot)
- 5 IR Line sensors (beneath the front edge of robot)
- 1 Electronic compass (accurate to 1 degree)
- 1 Beacon detector (capable of recognizing 15 different beacons)
- 1 Battery-level sensor
- 1 IR distance-measuring sensor (1-30 inches)
- 1 Servo controlled turret for the distance-measuring sensor
- Full motor control through `rForward` and `rTurn`

## 1.3 The Remote Robot's Responsibilities

It is important to realize that although RobotBASIC automatically sends out the appropriate data and properly utilizes the information returned, it is the robot's responsibility to actually carry out the requested actions and supply the necessary sensory data.

All of the robot's responsibilities will be handled by a control program written in C. The Pololu web page provides information on how to install a free C compiler on your PC and how to download the compiled files to the 3pi. Chapter 3 will examine the robot's new control program in detail. **It is important to realize that once this program has been written and installed on the robot, it never again has to be modified or downloaded because the robot itself will effectively become an extension of RobotBASIC.**

Before we can explore the robot's program though, we must construct the actual robot hardware, which will be examined in Chapter 2.

## 1.4 Summary

In this chapter, you have learned:

- ❏ About the Pololu 3pi robot and why it was chosen as the base platform for this project.
- ❏ How RobotBASIC's built-in wireless protocol communicates with external robots while maintaining compatibility with the integrated simulator.
- ❏ Which of RobotBASIC's simulated sensors will be implemented in the physical robot.

# Chapter 2

# New Hardware

The previous chapter gave you an overview of the Pololu project. This chapter will explore the hardware modifications that must be made to the standard 3pi in order to give it the sensory capabilities discussed in Chapter 1.

## 2.1 The Need for Multiplexing

The standard 3pi configuration reserves two I/O pins for serial communication, six pins to support the line sensors, and one pin to read the battery voltage. Several other pins are used for downloading programs to the 3pi, but most can be used for other purposes if we are careful. Normally, the 3pi supports three switches used for user input and a small LCD display. By giving up these features and using some of the programming pins, a total of seven I/O pins become available for our use. More information on these pins can be found at:

**http://www.pololu.com/docs/pdf/0J21/3pi.pdf**.

Seven pins might seem like a lot, but remember, we have to add five IR sensors and four bumpers not to mention a compass, a rotating distance sensor, and a beacon detector.

Normally, these additional sensors might require *at least* fourteen I/O pins.

One of the reasons the 3pi platform was chosen for this project was the way it reads its five line sensors. After studying the 3pi schematic (available on the Pololu web page), and examining the software provided with the 3pi, we determined that it was possible to read the new IR sensors and bumper switches through the I/O pins used for the existing line sensors. This *multiplexing* of three groups of data through one set of pins make it possible to interface all the desired sensors without adding additional microprocessor support.

All of the new sensors are available from Pololu or from Parallax (**http://www.parallax.com**). Refer to their web pages for complete specifications. Let's look at the actual sensors chosen for this project.

## 2.2 The IR Object Detection Sensors

Pololu offers numerous IR sensors. In order to detect objects around the front of the robot, we will use five Sharp GP2Y0D810Z0F sensors with a range of up to four inches. They provide a digital signal to indicate when an object is detected. The Pololu Item Number is 1134 and the sensor is shown in Figure 2.1.

**Figure 2.1**: This IR sensor is easy to mount and use.

## 2.3 Distance Measuring Sensor

The Sharp GP2Y0A21 distance measuring sensor is available from Pololu (Item Number 136). It measures distances from 4 to 32 inches and provides an analog interface.

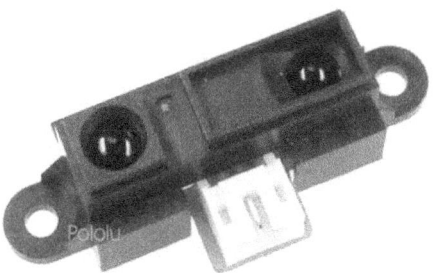

**Figure 2.2**: This distance sensor offers a simple analog interface.

This sensor will be mounted on a small servomotor (see Figure 2.3) so that it can be rotated to a desired angle to determine the distance to objects in front of the robot. The Pololu Item Number for the servomotor is 1053. Currently, it ships in a slightly different case. The important thing is that it is a subminiature servo with minimal weight and low current requirements.

**Figure 2.3**: An RC servo allows the distance sensor to rotate.

## 2.4 An Electronic Compass

An electronic compass allows a robot to make more accurate movements and even triangulate its current position (using beacons). The HMC6352 Compass Module from Parallax easily provides an accuracy of 1°, the same as the simulated robot. It is shown in Figure 2.4.

**Figure 2.4**: This Parallax compass module is accurate to better than 1°.

## 2.5 Beacon Detector

Later in the book you will see how to construct simple beacons that can be used for navigation purposes. The Vishay TSOP34156 IR detector module, available from Pololu (Item Number 837), will provide the ability to detect and identify 15 different beacons. It is shown in Figure 2.5.

**Figure 2.5**: This IR detector serves as a beacon sensor.

## 2.6 Bumpers

The robot will have four bumpers placed in the same relative positions as the simulated robot. Ideally, objects should be detected by the IR sensors, but as we will see later, such detection is not 100% reliable. The bumpers provide a fail-safe system for making sure the robot is aware of all collisions. Snap action switches, such as the one in Figure 2.6 from Pololu (Item Number 1403) will be used as the basis for the bumpers.

**Figure 2.6**: Snap action switches will be used as bumpers.

## 2.7 Bluetooth Communication

RobotBASIC will communicate with the 3pi robot through a Bluetooth link. On the robot side, we will use an Easy Bluetooth Module from Parallax (see Figure 2.7). We successfully tested other adapters, including Parallax's older EB500. You are not limited to Bluetooth adapters, though. Any wireless transceiver that operates through a virtual PC port should be acceptable.

**Figure 2.7**: This Parallax transceiver gives the robot the ability to communicate with RobotBASIC.

The PC end of the communication link also needs a Bluetooth transceiver. It seems reasonable, that since all Bluetooth devices meet industry standards, that compatibility should not be an issue. Unfortunately that is not the case, especially (based on our limited experience) with newer devices that have their own drivers (rather than the standard Windows' driver). Two PC Bluetooth dongles that we used, either would not connect or would not perform reliably with some robot-side adapters. Figure 2.8 shows the Abe dongle from Lego (**www.Lego.com**), and other sources, which worked flawlessly in all our tests. No doubt, there are many others that will work, but it is important to know that some compatibility issues exist.

**Figure 2.8**: The Abe Bluetooth PC dongle from Lego had no compatibility issues.

## 2.8 Physical Construction
Before we look at the schematic of the modifications, let's examine what our physical implementation will look like. Don't expect to duplicate every detail of our construction. Your board layouts and parts placement will certainly differ from ours. Ideally, sometime in the future, Pololu will consider making one or more circuit boards available to make these modifications easier.

## 2.9 Main 3pi Circuit Board Modifications
The standard 3pi circuit board has to be modified slightly in order to allow the multiplexing of the IR and Bumper sensors through the existing line sensor circuitry. The photo in Figure 2.9 shows the extra wires that connect the line sensors to a connector (to the second level) to carry signals to the multiplexing circuitry. These wires are connecting to the collector lead of the transistors in the line sensors. Looking at the board from the bottom, the collector lead is the lower-left corner of the 4-lead group for each line sensor. This is shown in Figure 2.9. When soldering these wires, use a low-wattage iron, pre-tint your

wires, and minimize the time you keep these joints hot in order to prevent possible damage to the sensors.

Some extra wires shown in Figure 2.10 were used to connect the standard ON/OFF switch to a new switch on the second level to provide a more convenient placement. The new wires were necessary on the prototype because the connector provided for this purpose was faulty on our 3pi.

**Figure 2.9**: The original line sensors must
be connected to the second level.

In our design, the connectors between levels serve as physical standoffs as well as electrical connections, making separation easier. Pololu sells expansion boards and extra connectors as shown in Figure 2-10 (Item Number 1039) to help meet your needs.

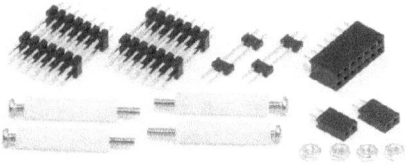

**Figure 2.10**: Electrical connectors can also serve as standoffs.

## 2.10 The Second Level Circuit Board

The bare second level circuit board (Pololu Item Number 978) is shown in Figure 2.11. It connects to the main level through three electrical standoffs as discussed above. One of these connectors carries the line sensor signals (from the new wires shown in Figure 2.9) and the other two carry the signals as shown in Figure 2.11.

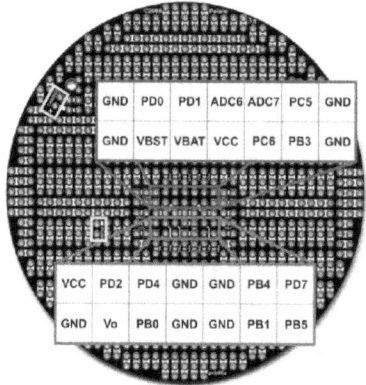

| GND | PD0 | PD1 | ADC6 | ADC7 | PC5 | GND |
| GND | VBST | VBAT | VCC | PC6 | PB3 | GND |

| VCC | PD2 | PD4 | GND | GND | PB4 | PD7 |
| GND | Vo | PB0 | GND | GND | PB1 | PB5 |

**Figure 2.11**: Connectors designed by Pololu carry the signals shown between the main board and the second level.

The second level holds the multiplexing circuitry as well as a Parallax Easy Bluetooth communication board, as shown in Figure 2.12. Also mounted on this level are the 5 infrared sensors. A new ON/OFF switch is located on the rear of this level to make it easy to reach.

**Figure 2.12**: The second level hold the Easy Bluetooth adapter, multiplexing circuitry, and the IR detectors.

## 2.11 The Third Level

A partially assembled third level is shown in Figure 2.13. It will connect to the second level much like the second connects to the first, with the electrical connections serving as physical standoffs. The third level was made from foam board and supports the switches that serve as the basis for the bumpers. As we will see later, the electrical standoffs will be soldered to a circuit board mounted on the bottom of level three.

Short pieces of rubber tubing (with an internal short piece of coat-hanger wire to ensure a proper curve) are crimped to the end of each switch lever. Notice there are six switches even though they will be electrically connected as four bumpers (to properly emulate the RobotBASIC simulated robot). The extra switches are necessary to keep the length of each "bumper" short to prevent potential binding.

Notice the servo and distance sensor also mounted on this partially constructed level.

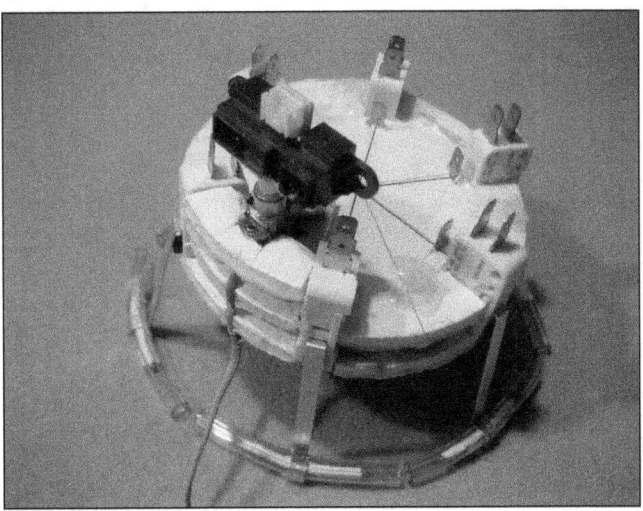

**Figure 2.13**: The third level is made from foam board.

Figure 2.14 shows how the beacon housing is added to the third layer. The outside housing is made from a fast-food salad dressing container. Notice the holes in the top of the container to allow wires to reach the compass, which will be mounted above the container

An interior slot in the beacon housing is made from balsa wood painted black to ensure that the beacon detector (mounted at the back of the slot) will only be struck by light entering directly forward of the slot. Later we added black foam to significantly reduce the size of the slit that allows light to enter. The final slot width was about 1/8 inch.

Figure 2.14 also shows a black cover (made from a coleslaw dish from a fast food restaurant) to give this level a more finished look. Later Figures will show the cover in place.

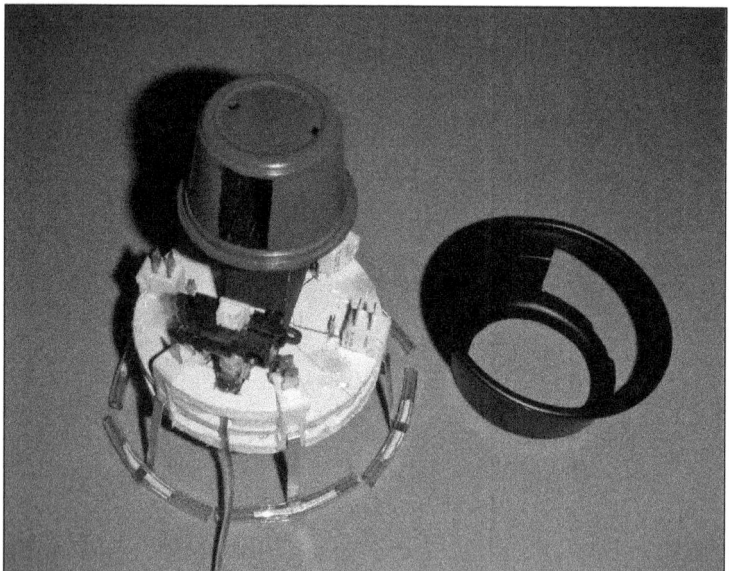

**Figure 2.14**: The beacon detector housing tops the third level, and eventually will support the compass.

A circuit board on the bottom of the third level (see Figure 2.15) allows a secure mount to the lower level. We used a ribbon cable to connect the upper sensors to the circuit board. Being able to totally disconnect this board made soldering much easier. Consider such things when you design your physical connections and layouts.

**Figure 2.15**: A circuit board provides support for the electrical standoffs.

## 2.12 Assembling the Levels

The three levels, when placed together create a nice looking robot. Notice that the compass is mounted on the very top of the robot, keeping it as far away as possible from the motor's magnetic fields. A black foam sheet serves as a skirt to improve the looks and hide the switches.

Once the housing is added the robot has a finished look as shown in Figure 2.17.

**Figure 2.16**: The robot is almost complete.

**Figure 2.17**: With the housing added the robot
has a more finished look.

Originally we planned to add a little weight to the backside of one of the levels, to increase the stability (since the stock 3pi has only one caster), but a small wire soldered under the front of the robot seems adequate. If extra weight is needed, consider adding a small second battery pack to power the new circuitry and improve overall battery life. The extra battery would be very helpful, as the current draw of the servomotor taxes the limits of many rechargeable AAA batteries.

## 2.13 Summary

In this chapter you have learned:

❑ About the physical hardware modifications that need to be made to the standard 3pi robot.

❑ How electrical signals are distributed between various levels.

❑ Why multiplexing will be used to minimize the number of port pins that will be needed to read all the required sensor data.

# Chapter 3

# The Electronics

In the last chapter, we examined the mechanical construction associated with the modified robot. In this chapter we will concentrate on the electrical modifications necessary to add the desired sensors to the robot.

As you know from Chapter 2, the IR and Bumper data will be read through the I/O pins used for the line sensors on a standard 3pi. In order to make this work, we will need some multiplexing circuitry to ensure that only one set of sensors is active at any one time.

## 3.1 Line Sensor Operation

If you look at the standard 3pi schematic (available on the Pololu web page) you will see that the standard line sensors have their own IR light sources mounted next to each light-sensitive transistor. If the light is reflected off a white surface, these transistors will turn ON and pull their respective I/O pin low. In the absence of light (in the case of a black line, for example) these transistors do not conduct and the I/O pin will float high.

Fortunately, the design of the 3pi already uses PC5 (I/O port C, pin 5) to control the light sources for the line sensors. Sending a zero to this pin effectively turns off the line sensors since the input lines simply float high when no light is provided. Chapter two showed how wires were soldered to the collectors of the line sensor transistors so that other sensors will be able to pull the I/O pins low, so that the new sensor data can be read through the same pins.

In order to make sure the standard line sensors still operate properly, we must make sure that the new sensors never pull the line sensor I/O pins low while the line sensors are active. Two primary ways of doing this are OC (open collector) circuits and tri-state circuits. Both of these methods allow the outputs being used to float to a high-impedance state as if the new circuitry is totally disconnected. We will utilize both methods in our design.

## 3.2 The Multiplexing Circuitry
Allowing multiple signals to be read through a single set of pins is called multiplexing. The circuit in Figure 3.1 shows one method for accomplishing our goal.

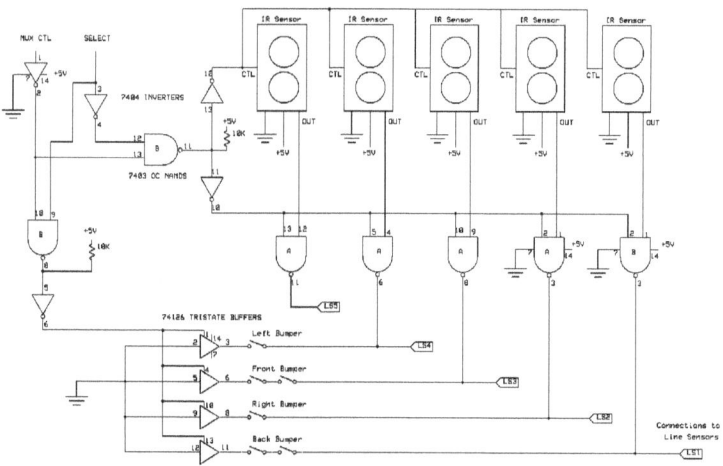

**Figure 3.1**: This multiplexing circuitry allows the IR and Bumper sensors to be read through the existing Line Sensor I/O pins.

## 3.3 IR Object Detectors
The five IR object detection sensors are shown in the upper right of Figure 3.1. The outputs from each of these sensors are gated through one input pin on each of five OC NAND

gates (7403 or equivalent). The outputs from the NAND gates connect appropriately to the line sensor I/O pins (refer back to Figure 2.9). LS1 refers to the robot's right-most line sensor while LS5 is on the robot's left.

The second input on each of these NAND gates is used for control. When this control line is HIGH, the gates will pass the IR data to the line sensor pins and when it is LOW, the IR sensors will effectively be disconnected from everything. We will examine how the control signal is generated shortly.

## 3.4 The Bumper Sensors

Notice, in Figure 3.1, that four tri-state buffers also connect to line sensor pins LS1 through LS4 through switches that serve as the bumper sensors. The line for LS5 is not used since the RobotBASIC simulation only has four bumpers.

The inputs to the tri-states are all grounded, so when the control input of the tri-states is HIGH (turning the tri-states on), one side of the bumper switches will effectively be tied to ground, and when the control line is LOW, the switches will not affect anything. This means, that when the bumpers switches are closed, that they can pull the line sensor pins low (to ground). Tri-state buffers and other miscellaneous parts can be purchased from Great Planes Electronics, DigiKey Corp., and JameCo Electronics.

We used **normally closed** switch contacts for the bumpers so that when the bumpers are not activated, the switches will pull the lines low. This is consistent with the line sensor logic, in that when the robot does NOT see a (black) line, the white surface will reflect light, turning on the phototransistors, pulling the inputs low. If you need to use normally open switches, it will be easy later to reverse the logic in software. Recall that the front and back bumpers are actually made from two switches each. These switches should be connected in series (as shown in Figure

3.1) so that pressing *either* of them will open the connection and allow the respective input pin to float high.

## 3.5 Controlling the Multiplexing Circuitry

The 7403 chip has four OC NAND gates in it, so we will need two chips because fives gates are needed as described earlier. That means we have three unused NAND gates to help build the control circuitry that will generate the signals that will enable/disable the multiplexing gates.

In order to enable the IR sensors, the NAND gate control line has to be HIGH, which means the inverter (7404 chip) driving those lines must have a LOW input. This LOW signal only occurs if the NAND gate driving it has both inputs HIGH, which only occurs if the MUX CTL and SELECT lines are both LOW.

Notice that each of the IR proximity sensors has a control line associated with it. While this line is not a standard pin on the sensors, the Pololu web page explains how this functionality can be added by cutting a trace and soldering one wire. While this modification reduces the power requirements slightly, it also prevents the test light on the IR sensors from indicating when objects have been detected. It is up to you to decide which of the above options is best for you. Notice how the IR sensors are turned on when the multiplexing NAND gates are enabled.

The control line for the tri-state buffers (for the bumper switches) must be HIGH to enable them, and that means the NAND gate (and inverter) creating that signal must have both inputs HIGH. This only happens when the MUX CTL is LOW and the SELECT line is HIGH. Let's see what this really means.

If the MUX CTL line is low, then both the bumpers and the IR sensors will be disabled, regardless of the value of the SELECT line. If the MUX CTL is HIGH, then the state of the SELECT line will determine *which* of the two new sensor systems is enabled (LOW enables the IR, HIGH enables the bumpers).

One thing should be mentioned. Since the NAND gates we are using are OC, they must have pull-up resistors (as shown in the schematic) to ensure that their outputs float HIGH when they are used as normal gates. We could have used a standard NAND gate chip for this purpose, but these are available to us without adding extra chips, and the resistors are easy to add.

Originally, we were planning on connecting the MUX CTL line to PC5 on the 3pi (port C, pin 5). This made sense, because the 3pi only pulls this line HIGH when it wants to turn on the lights for the line sensors. Some firmware conflicts exist (which will be discussed in Chapter 4) so we used PB4 instead. The SELECT line will be connected to PB5. Refer to the Pololu web page for more information on the 3pi I/O ports and Figure 2.11 to see from where these signals can be obtained when you are ready to start expanding your 3pi.

## 3.6 Interfacing the Remaining Sensors

We still have to interface the distance-measuring sensor (and its servo motor), the beacon detector, and the electronic compass to the main board I/O pins. The connections for these sensors are shown in Figure 3.2.

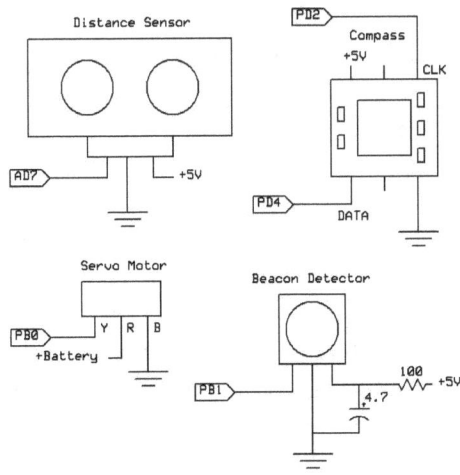

**Figure 3.2**: Five more I/O pins are required to interface the remaining sensors.

These sensors reflect how far the hobby robotics industry has come. Take the distance-measuring sensor, for example. Only a short time ago, you had only one option if you wanted such a device for your robot; you had to build it yourself. And doing so typically required oscillators, filters, amplifiers, perhaps even a phase-lock-loop. The same is true for each of the other sensors we are adding to the 3pi. Nowadays complete units are available and we only have to connect them to the robot's microcontroller.

## 3.7 The Distance-Measuring Sensor

Once we have power supplied to the distance-measuring sensor, it will, without any programs, without any additional circuitry, without any help from us at all, create an output voltage indicative of the distance being measured to objects within its range. And, since the 3pi has built-in hardware and software support for reading the value of analog voltages, a single one-wire connection is all that is needed to interface this sensor with the 3pi's microcontroller.

## 3.8 The Servomotor

The servomotor can be controlled just as easily. Once we connect it's control input to PB0 we can move the servomotor to precise angles simply by controlling the width of pulses being sent to it. By hot-gluing the distance measuring sensor to the servo's output bracket, our robot gains the ability to scan the area ahead and measure distances to objects that might block its path. As mentioned earlier, the servomotor draws, by far, the most current of any device that we have added to the stock 3pi. Adding additional batteries is recommended. At the very least, you should utilize the most powerful AAA batteries available. In our case, when we know the servo is not needed for the current project, we often disconnect the plug from the servomotor.

## 3.9 The Beacon Detector

The Beacon Detector is also a single chip that performs its function without the need for additional circuitry (except for a recommended filtering of the supply voltage as shown in Figure 3.2). The output of this chip normally is LOW, but will go HIGH when it sees an infrared signal pulsing at or near 56kh. Later in the book, we will discuss how to construct beacons that can be detected with this technology. For now though, we can interface the detector through I/O pin PB1.

## 3.10 The Compass

The compass too, is a complex circuit that would be time-consuming and expensive to construct if it was not available on a chip.

Only two I/O pins (PD2 and PD4) are needed to communicate with the compass. Later chapters will examine in detail the software needed for this communication.

The Parallax compass module we used has six pins and appears, at first glance, that it should fit in a standard DIP socket. Unfortunately, the spacing (from one side of the chip to the other) is slightly wider than a standard socket. This means you will need to make a socket from inline headers made for this purpose (see Figure 3.3). Simply break off two sections of three, and use them as the two sides of a socket.

**Figure 3.3**: Break off the needed length to create your own sockets.

## 3.11 Learning More About Interfacing

While this book explains how a variety of sensors can be interfaced to the 3pi hardware and utilized by the 3pi software, we realize that many readers may want a deeper understanding of such topics as well as information on other types of sensors, controlling motors and so forth. A book, *Hardware Interfacing with RobotBASIC* is being considered for publication. Please visit our web page for information and availability and do let us know your interest in such a book.

## 3.12 Summary

In this chapter you have learned:

- ❑ How to construct multiplexing circuitry that can interface our new sensors through the I/O lines used to read the 3pi's line sensors.
- ❑ How the beacon, compass, and distance measuring system are interfaced to the 3pi.

# Chapter 4

# The Primary 3pi Software

Chapter 3 explored interfacing the new sensors to the 3pi. In this chapter, we will examine the software needed to utilize those sensors.

## 4.1 Program Fragments

It is important to note that the figures in this chapter do not necessarily comprise the entire 3pi program. Also, the 3P source code is not necessarily in the order that it is presented here. Instead, the sections of code shown here are fragments of the program that make it easier to explain the functionality of the software.

The complete source code for the program can be downloaded from our web page (visit the HARDWARE tab) and then compiled using the software provided by Pololu. For your convenience, the compiled binary file is also provided on our web page if you would prefer to just download it to your 3pi (you will need a programming module, available from Pololu). See the Pololu documentation for details on compiling C programs and downloading binary files to the 3pi.

It is possible that improvements and error corrections may have been made to the downloadable file, so it might differ slightly from the code shown here. If modifications have been made, comments will be placed in the source code to explain the changes.

## 4.2: The main() **Routine**

A typical C program usually contains some #include statements and various subordinate routines and functions, as well as the establishment of any global variables, but the execution of a C program starts with the main() routine, so that is were our explanation will start too. It is shown in Figure 4.1.

```
int main(void)
    {
    initialize();
    while (1)
        BlueToothControl();
    }
```

**Figure 4.1**: Our 3pi program begins execution in the main() module.

An initialization routine is first called to setup the prerequisite conditions for what is to come. These actions include things like making sure I/O pins are properly setup for input or output, preparing the Easy Bluetooth module for communication, and initializing some sensors so they are ready to perform properly. Full details of this routine will be discussed later.

The next thing that happens in the main() routine is that an endless loop repeatedly calls the module BluetoothControl, which is responsible for nearly everything that happens in this program.

## 4.3 The BluetoothControl **Module**

The general function of this module is actually very straightforward. It deals with only three tasks as shown in the *simplified* version shown in Figure 4.2.

```
void BluetoothControl(void)
{
  // note: some variable initialization has
  //       been omitted here
  incoming[0]=0xff;
  incoming[1]=0xff;
  serial_receive_blocking(incoming,2,MyTimeOut);
  command=incoming[0];
  argument=incoming[1];
  if ((command==0xff) || (argument==0xff))
  {
    set_motors(0,0);
    return;
  }
  SensorBuffer[0]=0;
  SensorBuffer[1]=0;
  SensorBuffer[2]=0;
  SensorBuffer[3]=0;
  SensorBuffer[4]=0;
  switch (command)
  {
    case 3:  // rLocate command
      set_motors(0,0)
      for(i=0;i<30;i++) // do enough pulses
      {
        PORTB |= (1<<Servo);  // pulse line high
        delay_us(ServoCtr+ServoOffset);
        PORTB &= ~(1<<Servo); // return to low
        delay_us(20000-ServoCtr-ServoOffset);
      }
      OldServoAngle=0;
      play_from_program_space(welcome);
      delay_ms(500);
    break;
    // additional case statements follow
    // one for every command we will support
  }; // end of switch-case block

  // Here we will read the current data from
  // the bumper, IR, and line sensors and
  // return the information to RobotBASIC
}
```

**Figure 4.2**: This simplified module does most of the work necessary to control the 3pi from RobotBASIC.

**It is important to remember that the code shown in Figure 4.2 is NOT complete.** The 3pi program is a relatively large and complicated program and the goal here is to explain the principles associated with how the program works *not* to give you the entire program. Once you understand these principles, you can download the entire program from our web page and should find it straight forward to look through it and understand it's operation.

The first thing that happens in the `BluetoothControl` module is that both characters in the array `incoming[]` are initialized to `0xFF`. We will see why momentarily. Next, the Pololu-provided routine, `serial_receive_blocking`, is called to obtain two characters using the 3pi's serial port, which is connected to the Easy Bluetooth adapter. The parameters passed to this routine tell it to store the incoming data in the variable `incoming`, that there will be two characters, and to wait no longer than `MyTimeOut` (default 500 ms) for these characters to arrive. The two characters received are from RobotBASIC and should be the operation code and one additional parameter.

If for some reason both characters are not received within the allotted time, one or more of the characters will still be `0xFF`. If this happens, the module turns off both motors, using the Pololu-provided routine `set_motors`, and returns to the `main` program. Otherwise, this module assumes that it has a good command and an associated parameter that are stored in the variables `command` and `argument`.

> ☑ **Note:** RobotBASIC always sends out two bytes of data to the robot. If the second byte is not needed, a zero is sent.

The program then stores zero in all five bytes of the array `SensorBuffer[]`. These are the five bytes that the robot always returns to RobotBASIC. The function of the remaining code in the `BluetoothControl` module is to insert the appropriate data into this array (leaving the other bytes zero) before sending all five bytes out over the wireless link to RobotBASIC.

Next, the code checks to see which command has been received. This is accomplished by switching on the variable `command`, and then using multiple `case` statements to check for each of the possible values. In order to simplify the overview provided by Figure 4.2, only one case situation is shown. We will examine it, and many others shortly.

Following the `switch-case` construct, the program needs to determine the current status of three groups of time sensitive sensors (bumper, IR, and line). Let's see why the data from these sensors is time-sensitive.

Assume we are writing a RobotBASIC program to follow a line in a crowded environment. The program should tell the motors to move the robot slightly in some (appropriate) direction, and then check the values from the above three sensor groups to determine how the robot has moved in relation to the line being followed, and that the robot has not bumped into some obstacle blocking its path. RobotBASIC should NOT have to explicitly request such sensor information, so the data from these sensors is automatically sent back to RobotBASIC (and automatically used as needed) nearly every time data is exchanged between RobotBASIC and the external robot.

---

**Note:** Later in the book we will implement commands that give the user more control over this time-sensitive information. In particular, the user will be able to tell the 3pi that the line sensors are currently not needed, and the robot will then not take the time to determine their status and a zero will be returned in place of the actual sensor data.

---

The actual code for reading the sensors and returning the information will be examined shortly.

Now that you've seen an overview of how the `BluetoothControl` module works, we will examine each of the additional case-blocks that carry out tasks necessary to implement each particular command. We will start with the block for the `rLocate` command (`case 3`), which is shown in Figure 4.2.

## 4.4 Case 3: // rLocate

Normally, when writing a RobotBASIC program, one of the first robot commands that is issued is `rLocate`. In the simulation mode, this command creates the robot on the screen so that it is ready to receive and carry out additional commands.

In the case of the 3pi, this command will do several things as shown in Figure 4.2. First, the motors are both stopped to ensure the robot is initialized in a stationary position. Next, a `for` loop sends pulses to the servo motor (the number of pulses needed was determined experimentally). The purpose of these pulses is to cause the servomotor to move to its centered position.

Since this is our first real I/O operation, let's see how port bits on the 3pi are actually set and cleared. In this case, remember, that the servomotor is connected to Port B, Pin 0. We can set this bit using the statement

```
PORTB |= 0x01;
```

and clear the bit with

```
PORTB &= ~(0x01);
```

If a different bit is used, we have to substitute the proper hex number, and for some readers that might be confusing so we used an easier to follow approach. Our program initializes a variable equal to the actual bit number for each of our devices. The variable `Servo`, for example, has been set to 0 because we use pin 0 to control the servomotor. Likewise, the variable Beacon has been set to 1.

This means we can forget about creating the proper hex number ourselves, and let our program use shift commands to create the number for us. These two lines set and clear the servo bit.

```
PORTB |= (1<<Servo);
PORTB &= ~(1<<Servo);
```

In the case of the Beacon variable just mentioned, we need to read that bit instead of setting and clearing it. We can read that bit using the following expression in an if statement to see if it has a value of 1 or 0.

```
PINB & (1<<Beacon)
```

We will use these principles throughout our program to set, clear, and read port data. The Pololu AVR library now has some digital I/O routines, if you prefer to use them.

Let's return to the program in Figure 4.2. Notice how the servomotor port pin is *set* for a period of time, then *cleared* for another time period, using delay statements. In general, in order to command the servomotor to move its output shaft to a desired position over its possible range, we need to send it a HIGH pulse between 1000 and 1500 micro seconds, about 50 times per second. To that end, we have setup a variable called ServoCtr with the *centered* value of 1250.

Theoretically, if we set the Servo pin high, wait ServoCtr microseconds, set the Servo pin low, and wait approximately 1/50 of a second - and do this for at least a half second or so, the servomotor should move to its centered position. Since all motors are slightly different, the centering of your motor might require a slightly different value for ServoCtr.

If you are compiling your own code, you can substitute the value you need when the variable is initialized, but we offer a more general solution. We have a variable

`ServoOffset` that is initially zero, but can be changed from RobotBASIC to a value ranging from -128 to +127 (more on how RobotBASIC can change this value later). This allows users to easily change the way the Polulu 3pi works without having to modify any of the 3pi's internal code.

A major advantage of this method is that once your 3pi program is complete and downloaded to the robot, that it never again has to be tampered with. As you will see later, many minor adjustments to the 3pi's internal operation will be possible directly from a RobotBASIC application.

Look again at the code in Figure 4.2. It should be obvious now how the servomotor is moved to its initial centered position. Notice also that the variable `OldServoAngle` has been set to zero. As we will see later, this variable will allow us to only move the servomotor when a newly requested angle differs from its current position. This methodology greatly conserves battery life because the servo only draws current when we need to move it.

Next, the `rLocate` code will play a short sequence of notes on the 3pi speaker to let the user know it has been initialized. Finally, a short delay is invoked, and this section of code is complete.

The following sections of this chapter will provide explanations for the `case`-code necessary to handle each of the major RobotBASIC commands.

## 4.5 Case 6: // rForward

When the robot receives the command 6, it knows the argument byte contains the number of units it should move forward. The code to handle this operation is shown in Figure 4.3.

There are several conditions that this code must handle. First it checks to see if the robot is being asked to only move one unit, which is the typical case when the robot is following a line or hugging a wall, or any operation where it must make sure it does not bump into an object. We could have the robot move a very small distance and then stop the motors, so that it is assured that the robot does not move very far before RobotBASIC acquires new sensory data. In fact, this is what the simulation does.

```
case 6:
  if (argument==1)
    set_motors(speed+Loffset,speed+Roffset);
  if (argument==0)
    set_motors(0,0);
  if (argument>1)
    {
    if ((argument<50) || !CompassEnabled)
      {
      set_motors(speed+Loffset,speed+Roffset);
      delay_ms(MoveTime*argument);
      set_motors(0,0);
      }
    else
      {
      temp=Read_Compass();
      set_motors(60,60);
      delay_ms(15);
      temp1=0;
      for(i=0;i<(argument*MoveTime)/96;i++)
        {
        set_motors(speed+Loffset,speed+Roffset+temp1);
        comp=Read_Compass();
        temp2=abs(temp-comp);
        temp3=temp2;
        if (temp2>50)
          temp3 = 360-temp2;
        if(temp!=comp)
          {
          if( ((comp<temp)&&(temp2<50)) || ((comp>temp)&&(temp2>50)) )
            temp1=-(temp3);
          else
            temp1=temp3;
          }
        }
      }
    }
  break;
```

**Figure 4.3**: This code handles moving the robot forward.

If we have the robot do this though, it will move in a very jerky way, because of the starts and stops intertwined with the commands and data being sent and received over the wireless link. In order to prevent this jerky motion, we will set the motor speed to a moderate rate (the default value for the speed is 40) and just let the motors run until the next command is received. Since Bluetooth allows us to communicate with the robot approximately 10-15 times per second, as long as the motor rate is reasonably slow, the robot will not move far in relationship to the last sensor readings.

The next special case that has to be handled is when the distance to be moved is zero. Normally, we would not expect RobotBASIC programs to issue an rForward 0 command since the robot does not move unless commanded to do so. This command is useful when dealing with the real robot though, because it gives us an easy and quick way to halt the robot. This is accomplished by turning off both motors as shown in Figure 4.3. Even without this command, the robot should normally stop all motion in 500 ms (if it does not receive a command in that time period) due to the timeout period used at the beginning of Figure 4.2.

The next if statement checks for the final special condition - that the robot is being commanded to move some distance on its own (any distance greater than 1 unit). In this situation there are two possible ways to perform the move. The easiest of these is to set both motors to an appropriate speed, and then wait a time proportional to the desired distance. If we calibrate the speed and wait-time properly, then the real robot should move similarly to the RobotBASIC simulation. Since the default radius of the simulated robot is 20 pixels, an rForward 40 moves the simulation a distance equal to the robot's diameter. Ideally, an rForward 40 command should also move the 3pi a distance equal to its diameter, which is about five inches including the bumpers.

Once the motors are turned on, the program simply delays a short time to allow the robot to move. The delay period needs to be proportional to the desired distance to be moved. It was experimentally determined that a delay equal to the argument * 24 would allow the 3pi to move a distance similar to the simulation. The variable MoveTime is initialized to 24 and used as shown in Figure 4.3.

It is important to realize that 24 was appropriate for our robot, but that does not mean it is right for yours due to physical differences involving motor efficiency, friction, etc. You can, of course, substitute a proper value for MoveTime when you compile your code, but, as we will see later, you will be able to alter the value of MoveTime from your RobotBASIC applications.

Unfortunately, the minor variations in motor efficiency and friction discussed above, can also make your robot drift slightly to the right or left instead of moving in a straight line. Notice the left and right offset variables (Loffset and Roffset) in the code of Figure 4.3. The value of these variables are normally zero, but they too can be modified from RobotBASIC. Assigning them a small positive or negative number can help compensate for any drift that occurs with *your* robot.

There is another way to help ensure that the robot moves in a straight line, and that is to let it use the electronic compass to automatically compensate for drift. We need some way of letting the user decide which of these two methodologies to use.

Our 3pi program will utilize a variable called Compass-Enabled (default disabled) to help decide if the compass should be used. In later chapters we will see how the value of this variable can be altered by a RobotBASIC program.

Notice the next if-statement in Figure 4.3. If the compass is not enabled, or if the robot is moving less than 50 units (an arbitrary distance) then the motors are simply turned on for a time proportional to the parameter passed from RobotBASIC as the argument of the rForward command. In this case, the values of Loffset and Roffset are used to make the robot move in a relatively straight line.

If the compass is enabled, and the distance being moved is relatively long, then the else-block of the previously mentioned if-statement performs the move with compass compensation. Lets see how that works.

The first thing that happens is the Read_Compass() function stores the robot's current heading in the variable temp (Chapter 6 will provide details on how the compass operates). Next the motors are started and then a for-loop will repeat a series of actions a number of times proportional to the desired distance to be moved.

Again, the variable MoveTime tries to calibrate the movement to ensure it is similar to that of the simulated robot. The number of times through the for-loop is further adjusted by dividing it by 96 (also chosen experimentally based on the time taken to read the compass, etcetera, inside the for-loop).

Inside the for-loop an if-statement checks to see if the robot's current heading is equal to the original heading (temp). If the two headings are not equal, a several if-statements decide how to change the value of temp1 to decrease or increase the speed of the right wheel.

Basically, the speed of the right wheel will be altered proportional to the error. For example, if the original heading was 10°, and the robot is currently heading 12°, then the right wheel speed will be *increased* by 2 (12-10). If the current heading was 7, then the right wheel speed would be d*ecreased* by 3 (7-10). This is an easy principle to grasp, so you might wonder why the code to implement it is so cryptic. The extra complexities simply handle situations that arise when the robot's heading is close to due North. For example, if the robot's original heading is 2° and its current heading is 3° to the left, the compass will report the heading as 359°. Simply subtracting the values would cause a speed correction of 357 instead of 3. The extra code in Figure 4.4 corrects such situations.

## 4.6 Case 7: // reverse

An operation code of 7 is generated by RobotBASIC when an rForward command is executed with a negative distance specified. You might wonder why RobotBASIC does not just use one op-code and simply send a ± argument. Since only one byte is used for this parameter the distance would have been limited to +127 or –128. By using two op-codes and passing only positive numbers, the range for both forward and reverse is enlarged to 255. This same philosophy is used for rTurn.

The code we used for moving backwards uses the same methodology as that shown in Figure 4.3 for forward, *except*, that the compass is never used to compensate for drift when the robot moves backwards. This decision was made because in all the algorithms we have developed, the robot never moved backward for any significant distance. If you feel different about this decision, it should be easy for you to adjust the code for the backward operation.

## 4.7 Case 12: // rTurn (right)

When an op-code of 12 is received, a right-turn is indicated and the variable argument indicates the number of degrees to turn. The value of this variable controls how this code performs. The first if-statement, for example, checks to see if the robot is being asked to turn only one degree, a typical request for many situations.

When developing algorithms such as following a line or hugging a wall, for example, the robot will often be asked to turn only one degree (the smallest angle RobotBASIC can request) before the sensors are checked to decide on the next action.

We could use the compass to ensure that the robot turns one degree, but as we will see shortly, that can be time consuming because the robot can overshoot the desired position and have to turn back slightly. For sensory-based algorithms the exact amount of the turn is not important;

not as long as the turn is small enough to ensure that vital sensory positions are not missed. To that end, we need only to make the robot turn a small amount.

The RobotBASIC simulated robot always turns by rotating around its center. This is done by turning one motor forward, while the other is reversed. While this is perfectly acceptable for the simulator, there are advantages for a real robot to have choices as to how it makes its turns. We will give the 3pi a variety of turn styles and we will use the variable TurnStyle to indicate which type of turn to use.

If TurnStyle is 0 (the default), then the robot will rotate around its center just like the simulator (the motors are both turned on, at the same speed, but in opposite directions). If TurnStyle is 1 or greater, one of the motors will be turned on at the normal speed and the other motor will be turned on, also in a forward direction, but at a slower rate, thus generating a slower turn when compared to a rotation. The value of TurnStyle can be anything from 1 to 8. Since we are multiplying the normal speed (including offset corrections) by TurnStyle, and then dividing by ten, we are effectively setting the slow wheel speed to a percentage of the faster wheel. For example, if TurnStyle is 4, the slower wheel speed will be 40% of the faster wheel. The higher the value of TurnStyle, the slower the robot will turn. Future chapters will demonstrate how to use this feature.

If a robot moves using only rotational turns a jerky motion is inevitable. A jerky motion occurs in the simulation too, but the robot responds so quickly (because the simulation has no inertia) that it is seldom noticed. The use of these turn styles can minimize and even eliminate the jerky movement created when a real robot turns only by rotation.

Look at the code in Figure 4.4. When the robot is asked to turn only one degree, two if-statements determine how to react. If the current `TurnStyle` is zero, the motors activate a standard rotational turn. If `TurnStyle` is larger than zero, the slow motor's speed will drop to between 10% and 80% of its normal speed (the value of `TurnStyle` is limited elsewhere to 1 to 8). Later chapters will explain how RobotBASIC applications can alter the value of `TurnStyle`.

> ☑ **Note:** While the overall motion of the 3pi is relatively smooth, large (man-sized robots) or even small but fast robots might benefit from additional technologies. For such situations, you might consider speeding up the communication process between RobotBASIC and the micro-controller handling the low-level processes by using more expensive wireless transceivers or even embedding the PC in the remote robot with a wired link to an advanced micro-controller like Parallax's Propeller-based Robot Controller. Such combinations could add many features including the ability to transition from one motor speed to the next (using a ramping algorithm), while continuing to monitor and react quickly to the time-sensitive sensors.

Refer again to Figure 4.4. The next `if`-statement checks to see if the robot has been asked to turn zero degrees. When this happens, the motors are commanded to stop.

The next `if`-statement handles all other situations by checking to see if the argument is greater than one. If it is, another `if` decides if the argument is less than 15 or not. When the angle is large (greater than 20°) the robot will use the compass (if it is enabled) to move the 3pi to the specified angle. Let's see how that is done.

First, we read the compass to find out the current robot heading. Then we add the desired turning angle (the argument) to get a new heading. We make sure the new heading is in the range of 0-359 and then use another routine (MoveToAngle) to actually move the robot to the desired position. We will discuss MoveToAngle in Chapter 5.

```
case 12:
  if (argument==1)
    {
    if (TurnStyle==0)
      set_motors(speed/2,-speed/2);
    if (TurnStyle>=1)
      set_motors(speed+Loffset,(TurnStyle*(speed+Roffset))/10);
    }
  if (argument==0)
    set_motors(0,0);
  if (argument>1)
    {
    if ((argument<20) || !CompassEnabled)
      {
      set_motors((speed*2)/3,-(speed*2)/3);
      delay_ms(RotTime*argument);
      set_motors(0,0);
      }
    else
      {
      temp=Read_Compass();
      temp+=argument;
      if (temp>=360) temp-=360;
        MoveToAngle(temp,speed);
      }
    }
  break;
```

**Figure 4.4:** This code turns the robot to the right.

If the new angle is less than 20° (even if the compass is enabled), the robot simply rotates at a slow speed (half of the normal speed) for a time proportional to the amount of turn required. The turn-time is controlled by the value of the variable RotTime, which was experimentally set to 12. The value of this variable maybe changed from RobotBASIC (more on this later) so that you can calibrate the angular movement for your robot so it is reasonably correct.

It is important to realize here, that the turn styles mentioned earlier only apply to turn requests of exactly one degree. All other requests result in a rotation just like the simulation. Because of this, the real robot responds almost exactly like the simulation, even when alternate turn styles are enabled. Applications later in the text will examine this principle in more detail.

The code for turning left (`case 13:`) is not shown, as the logic for it is identical to the right turn algorithm except that the motors directions are reversed.

## 4.8 Case 96: // rBeacon

Before we can understand how to recognize a beacon, we must understand what a beacon is and how it operates. Later chapters will examine how to build a beacon. For now, let's just explore their functionality.

The beacon-detector chip we are using is designed to react to an IR frequency of 56kh, which means all our beacons must be oscillating at that frequency. In order to create 15 different beacons, we will have each beacon periodically turn off for a short time as shown in Figure 4.5.

Each beacon will generate a 56kh infrared signal for a repeating period of approximately 10ms. This time is indicated in the Figure as the ON TIME. This time period is not critical, but something close to 10ms is recommended.

Each beacon will have a different OFF TIME. Beacon #1 is off for 100 microseconds, Beacon #2 is off for 200 microseconds, etcetera. This means the maximum off time is 1.5ms for Beacon #15.

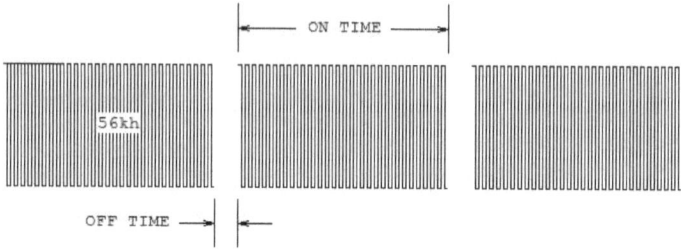

**Figure 4.5**: Each beacon generates this waveform.

Let's see how the robot can determine if a beacon has been detected and if it is the correct one. Normally, in the simulation mode, a numeric argument, indicating the color of the beacon, is passed to the rBeacon() function. When using the 3pi, we will still pass the numeric argument, but beacon requested will be based on the off-times given above. The function will return TRUE if the beacon detected matches the one requested. If the user requests Beacon #0, then any beacon will cause TRUE to be returned. The code in Figure 4.6 shows how this can be accomplished.

When rBeacon is used in a RobotBASIC program it will send out the code 96 followed by the Beacon number provided, which is why Figure 4.6 uses the case 96. Our beacon detector chip outputs a low or zero when it detects a 56kh signal, so the first line of code checks for a high, which would mean that either no beacon is in sight, or if there is a beacon, it is currently in its off-time. Since the off-time is always less than 1.5ms, we will wait 2ms and try again.

At this point, if the signal is still high then no beacon has been found. If one has been detected, and if a non-zero argument indicates a specific beacon is requested, then we must measure the off time to see if this is the desired beacon.

In order to measure the off-time, the code in Figure 4.6 must first wait for the data line to fall low, which is done with the empty `while`-loop. Since all the time periods are in multiples of 100 microseconds, we wait 50 microseconds to place us in the middle of the first period. Next we enter a `for`-loop to count the number of periods the data line stays low. Each pass through the loop starts with a 100 microsecond delay to move the time to the middle of the next period, where the data line is tested.

```
case 96:
   if (PINB&(1<<Beacon))     // it it happens to be High now
      delay_ms(2);           // wait and try again
   if (!(PINB&(1<<Beacon)))
      {
      // beacon has been detected
      if (argument>0)
         {
         // see which one it is
         // first wait till beacon signal stops
         while (!(PINB&(1<<Beacon)));
         // then count the time till it starts again
         delay_us(50);
         for (i=1;i<=15;i++)
            {
            delay_us(100);
            if (!(PINB&(1<<Beacon)))
               break;
            }
         if (i==argument)
            {
            SensorBuffer[4]=1; // true (found it)
            }
         }
      else
         {
         SensorBuffer[4]=1; // return True for ANY beacon
         }
      }
   break;
```

**Figure 4.6:** This code determines if a requested beacon is found.

When the data line is found to be low a `break` statement terminates the loop early and the variable `i` represents the beacon number. An `if`-statement is used to see if the beacon found matches the one requested. If it is, the proper element in the `SensorBuffer` array is set to one to indicate the beacon was found. This also happens if any beacon was detected and the argument passed from RobotBASIC is zero.

## 4.9 Case 24: // rCompass

The case block that handles the compass command is easy to code because special functions have been created to handle all the details of interfacing with the hardware. Because of its complexity, those functions will be covered in a later chapter. For now, Figure 4.7 shows how easy it is to pass the data back to RobotBASIC.

```
case 24:
  sum=Read_Compass();
  SensorBuffer[3]=sum>>8;
  SensorBuffer[4]=sum&255;
break;
```

**Figure 4.7**: This code provides the compass heading.

The first line in the block uses the Read_Compass() function to obtain the current compass angle, which is a 2-byte number. The two bytes are placed in the appropriate position of the SensorBuffer array.

## 4.10 Case 192: // rRange (right)

The rRange() function in RobotBASIC returns the distance measured by the distance sensor. The 3pi variable DistMode will control whether the distance provided is in inches, pixels, or the raw value provided from the sensor (allowing manipulation by RobotBASIC to achieve maximum resolution when needed). We will see later how RobotBASIC can set the value for DistMode.

The distance sensor produces a nonlinear analog signal as its output. This is easily read using the standard 3pi hardware and library functions. Because of the non-linearity of the signal, we will use an array (serving as a look-up table) to translate the raw data to pixels and then to inches. The numeric value of the raw data ranges from 0 to 159 or so, with larger numbers representing shorter distances. If we used the raw data as an index to an array, it would take a 160 element array to accurately translate to a pixel measurement. Fortunately, a general idea of the distance being measured is all that is usually needed for most robotic algorithms. Because of that, and in order to minimize memory requirements, a translation array of only 32 elements will be used as shown below.

```
static unsigned char translateDist[32]={255,255,255,
    240,192,155,128,104,88,75,68,60,52,46,42,38,35,32,
    29,26,24,22,20,18,16,14,12,11,10,9,8,0};
```

Since we only have 32 elements in the array, we cannot use the raw distance (0-160) as the index to the array. The raw distance will be divided by 5 so that it has a range of 0-31 instead of 0-159. This number will be used as an index into the array and return the number of pixels representing that distance (the numbers in the array were found experimentally). Based on the size of the 3pi in relation to the simulated robot, an inch is roughly equal to 8 pixels.

In addition to all of the above, the code that handles the ranging function must also control the servomotor that can point the distance sensor in the desired direction. As with previous examples, there are two blocks supporting the rRange command, one for servo angles to the left and one for angles to the right.

The code to implement all of the above (for servomotor angles to the right) is shown in Figure 4.8. The variable argument holds the number of degrees to be moved. We don't want to take the time, or utilize the power, to move the servomotor if it is not necessary, so the first line of the code checks to see if the requested angle is different from the last known position (OldServoAngle). This variable will be discussed more later.

If the requested angle is different, then a for-loop sends 30 pulses to the servomotor (enough to ensure it will move to the desired position). The width of these pulses (not the number of pulses) will determine the final position of the motor.

Inside the for-loop of Figure 4.8, the servo's control line is brought high for a period based on the following formula.

$$ServoCtr-(10*argument)+ServoOffset$$

As you recall, the variable ServoCtr holds the time delay needed to move the motor to a center position. Since this position might be slightly different for different servomotors, we add to this the value of ServoOffset. If we waited this amount of time, and dropped the motor's control line low again (and did this 20 or 30 times) the motor would move to its centered position.

```
case 192:
    if (OldServoAngle!=argument)
        {
        for(i=0;i<30;i++) // do 30 pulses
            {
            PORTB |= (1<<Servo);  // bring control line high
            temp = ServoCtr-(10*argument)+ServoOffset;
            delay_us(temp);
            PORTB &= ~(1<<Servo); // return control line low
            delay_us(20000-temp);
            }
        OldServoAngle=argument;
        delay_ms(20);
        }
    // now take reading
    set_analog_mode(MODE_8_BIT); // 8-bit conversions
    sum = 0;
    ave = 0;
    for(samples=0;samples<10;samples++)
        {
        start_analog_conversion(TRIMPOT); // start conversion
        while (analog_is_converting())
            ; // wait till conversion is done...
        sum += analog_conversion_result(); // get result
        }
    ave = (sum / 10); // compute 10-sample average
    if (DistMode==2)  // raw
        SensorBuffer[4]=ave;
    if (DistMode==1)  // inches
        {
        ave=ave/10; // reduce resolution to 1/16
        if (ave>15)
            ave=15;
        SensorBuffer[4]=TranslateDist[ave]/4;
        }
    if (DistMode==0) // pixels
        {
        ave=ave/10;
        if (ave>15)
        ave=15;
        SensorBuffer[4]=TranslateDist[ave];
        }
    break;
```

**Figure 4.8**: This code points the distance-measuring sensor in the proper direction and then takes a reading.

If we want the motor to move to a different angle, we can increase or decrease the delay (decreasing it moves our motor to the right). Since argument specifies an angle to move we need a formula to translate the angle into an appropriate time delay. It turns out, for the servo we used, that each 10 microseconds of delay moves the motor about 1°. The final thing that happens in the loop is to delay a relatively long time to ensure that the servo motors receives about 50 control pulses each second. Remember, this for-loop only executes if the servomotor is NOT already pointing in the proper direction.

Since the motor has just been moved to a new angle, the variable OldServoAngle is set to the current angle. Because of this, future calls to this routine will only move the motor when necessary. Finally, a short delay lets the motor settle before a distance reading is taken.

At this point in the code of Figure 4.8 we are ready to make the reading. Actually, we will use another for-loop to take ten 8-bit readings and average them together to improve on the accuracy. Taking the reading is really easy, because the standard 3pi has a small potentiometer (variable resistor) mounted on its underside that delivers a variable voltage (to Pin AD7), and Pololu provides all the necessary commands and functions to determine the value of this voltage.

Luckily, Pololu had the forethought to provide a jumper that allows the potentiometer to be disconnected from the Pin. This means we can use Pololu's standard trimpot routines to read the distance signal (because we have the output from the distance sensor connected to AD7 as discussed in Chapter 2).

Once we have the raw distance value (stored in the variable ave) we must decide how to use it based on the value of DistMode. When DistMode is 2, we simply place the raw data into the proper position of the SensorBuffer array, so that it can be sent back to RobotBASIC (more on this shortly).

If `DistMode` is zero, the raw distance data needs to be converted to pixel information. This is done by dividing the raw data by 10 and ensuring that it ranges from 0-15 so that it can be used as an index into the `TranslateDistance` array mentioned earlier. This lowers the resolution considerably (we only get 32 different values instead of 160), but in practice it works reasonably well, especially since the non-linearity of the distance sensor gives more readings when objects are close to the robot.

If `DistMode` is 1, the distance should be returned in inches. This is easy because an inch roughly translates to 8 pixels based on the relative sizes of the 3pi and the simulated robot. Therefore, the code for this operation is just like the code for pixels, except the final answer is divided by 8.

## 4.11 Case 193: // rRange (left)

The code for this section is basically the same code for **rRange** to the right (Section 4.10). The only real difference is that the pulse sent to the servo has to be *increased* in width in order to make the motor turn left.

## 4.12 Case 108: //Battery level

Just as Pololu has provided routines to read the analog voltage from a trimpot as discussed in Section 4.10, they have also provided routines to read the battery voltage, giving us an easy way to determine when the batteries need charging (particularly nice if you wish to make your robot autonomously locate and utilize a charging station).

The simulated robot in RobotBASIC can determine the percentage of charge left in the battery using the function `rChargeLevel()`. Figure 4.9 shows the 3pi code that will allow the 3pi to provide proper values for this function.

Since the voltage of a fully charged (rechargeable) battery is about 5200 millivolts we can get a percentage of charge simply by dividing that number by 52. In our 3pi proto-type, the battery became too low to operate when this number dropped to 70% or so, but your robot will probably differ.

```
case 108:
   battery=read_battery_millivolts();
   battery=battery/52;
   if (battery>100)
      battery=100;
   SensorBuffer[3]=battery>>8;
   SensorBuffer[4]=battery&255;
break;
```

**Figure 4.9**: This code determines the current battery condition.

If you prefer, you could just return the millivolt reading and let your RobotBASIC programs handle the calculations. If you do this though, simulator programs will have to be modified slightly to make them run properly with the real robot. In case you wish to do this, the code in Figure 4.9 shows how to return larger two-byte numbers such as the battery voltage. Since our code returns a maximum value of 100 (which easily fits in one byte), the extra line of code shown in the Figure is unnecessary (although it works fine).

## 4.13 Dealing with the rSpeed Command

The rSpeed command is used to add delays to the simulated robot's movements and does not translate well for controlling the speed of the real robot. Because of that we will simply ignore any rSpeed commands sent by RobotBASIC to the 3pi, as shown in Figure 4.10. Chapter 5 will provide a method for controlling the 3pi's speed.

```
case 36:
   // ignore rSpeed commands
break;
```

**Figure 4.10**: Simulator speed commands are ignored.

## 4.14 Reading the Time-Sensitive Sensors

At the end of the `switch-case` construct discussed through out this chapter, we need to find the values for the Line, IR, and Bumper sensors, place them in the first three bytes of the `SensorBuffer`, and then return the `SensorBuffer` to RobotBASIC (refer to Figure 4.2). Remember, RobotBASIC does not have to ask for this information. The 3pi will gather and return it almost every time communication occurs between RobotBASIC and the remote robot (there are only a few *possible* exceptions, which will be discussed later). The code to perform these actions is shown in Figure 4.11.

While the code in Figure 4.11 is logically correct and fully functional, it may differ slightly from the full program on our web page in order to improve the readability by making it fit on a single page.

If you examine Figure 4.11 you will see there are three primary `if`-blocks that handle the reading of Line, IR, and Bumper sensors. In general, each of these blocks must read the associated sensors and store the properly formatted data into an appropriate byte of the `SensorBuffer`.

The beginning of each block is responsible for sending the appropriate control signals to the multiplexing circuitry (discussed in Chapter 2). When reading the Line Sensors, for example, we must make sure the multiplexing circuit disconnects the Bumpers and IR Sensors from the I/O ports. Conversely, when reading the Bumpers or IR Sensors, we must turn OFF the lower Lights for the Line Sensors (effectively eliminating the Line Sensors as inputs) and enable the desired sensors through control of the multiplexing circuitry.

In all cases, once the proper sensors have been selected, the Pololu-provided function

```
read_line_sensors_calibrated()
```

can be used to read the values (automatically scaled between 0-1023) from the selected sensors into an array (in this case, SensorValues[ ]). A series of if-statements can then be used to see if the current values are above or below a preset limit (HiLevel). If they are above, then that bit-position is set to a 1, otherwise it is assumed to be 0.

```
if (LineSensorsEnabled) // OK to read Line Sensors
  {
  PORTB |= (1<<4); // line high to disable IR and bumpers
  delay_ms(3); // required for reliable reads
  read_line_sensors_calibrated(sensor_values,IR_EMITTERS_ON);
  sv0=sv1=sv2=sv3=sv4=0; // zero all bit positions
  if(sensor_values[4] > HiLevel) sv4 = 8;
  if(sensor_values[3] > HiLevel) sv3 = 1;
  if(sensor_values[2] > HiLevel) sv2 = 2;
  if(sensor_values[1] > HiLevel) sv1 = 4;
  if(sensor_values[0] > HiLevel) sv0 = 16;
  svtotal = sv4+sv3+sv2+sv1+sv0;
  SensorBuffer[2]=svtotal; // prepare to send sensors back
  }
// Now read the IR sensors
  PORTC &= ~(1<<5);      // turn off Line Sensors lights
  PORTB &= ~(1<<IRctl);  // set low to enable IR sensors
  PORTB &= ~(1<<4);      // enable extended sensors
  delay_ms(5); // required for reliable reads
  read_line_sensors_calibrated(sensor_values,IR_EMITTERS_OFF);
  PORTB |= (1<<4); // disable extended sensors
  sv0=sv1=sv2=sv3=sv4=0; zero all bit positions
  if(sensor_values[4] > HiLevel) sv4 = 1;
  if(sensor_values[3] > HiLevel) sv3 = 2;
  if(sensor_values[2] > HiLevel) sv2 = 4;
  if(sensor_values[1] > HiLevel) sv1 = 8;
  if(sensor_values[0] > HiLevel) sv0 = 16;
  svtotal = sv4+sv3+sv2+sv1+sv0;
  SensorBuffer[1]=svtotal;
// now read the Bumper sensors
  PORTC &= ~(1<<IRctl); // off Line Sensor lights
  PORTB |= (1<<IRctl);  // lines high to enable bumpers
  PORTB &= ~(1<<4);     // enable extended sensors
  delay_ms(1);
  read_line_sensors_calibrated(sensor_values,IR_EMITTERS_OFF);
  sv0=sv1=sv2=sv3=sv4=0; // zero all bit positions
  if(sensor_values[4] > HiLevel) sv4 = 1;
  if(sensor_values[3] > HiLevel) sv3 = 2;
  if(sensor_values[2] > HiLevel) sv2 = 4;
  if(sensor_values[1] > HiLevel) sv1 = 8;
  svtotal = sv4+sv3+sv2+sv1;
  SensorBuffer[0]=svtotal; // prepare to send sensors back
// and send back the sensor data
  serial_send_blocking(SensorBuffer,5);
  }
```

**Figure 4.11**: Line, IR, and Bumper data is nearly always sent back.

All of the bit-values are then combined to create a single number representing that sensor group. This number is eventually placed into the proper position in the SensorBuffer.

The final line in Figure 4.11 utilizes a Pololu-provided function to send all five bytes of the `SensorBuffer` back to RobotBASIC using the 3pi serial port, which is connected to the EasyBlueTooth transceiver. This is an important concept that deserves further attention before ending this chapter.

Some of the code fragments discussed in this chapter (such as the beacon detector and compass modules) generate sensory data that is needed by a RobotBASIC application. These modules only need to place the data into the appropriate position in the `SensorBuffer`, which is automatically sent to RobotBASIC as shown in Figure 4.11.

The program stored in the 3pi robot does not need to do anything except send the data as described. RobotBASIC (when enabled with the `rCommPort` command) will automatically capture this data and store it in its own buffer system and utilize it appropriately when the current application needs or asks for it. This is a unique concept that makes it possible to build powerful robotic hardware that can be inspected and controlled by novice programmers utilizing the extremely easy-to-use RobotBASIC language. Furthermore, students and hobbyists can use the simulator to learn how to deal with a real robot even before they have one.

## 4.15 Initialization

All the code in this Chapter assumes that proper initialization has been provided, as discussed in Section 4.2. Figure 4.12 shows the code that provides this for us.

A Pololu provided routine is first called to perform the general initialization required by the 3pi. Next, the 3pi's I/O ports must be set up properly.

Each individual I/O pin on the ports of the 3pi can be individually and independently configured as an input or an output pin. Each port has an associated data direction register (DDR). The bits in the DDR control whether that pin position is an input or output. Placing a 1 in a bit position identifies that pin as an output and a 0 indicates input.

Many of the I/O pins on the 3pi are being used in the same I/O state as a stock 3pi, so only a few DDR pins need to be modified (including two pins in Port B and one in Port D). See Figure 4.12 for details.

Furthermore, both the data pins driving the servo and multiplexing control need to be set low for their initial state.

At this point in the routine, the I/O pins associated with the compass are normally initialized. The particulars have been omitted here though, as the compass code is discussed in detail in Chapter 6.

Next, the routines used to read the line sensors (and the IR and bumper sensors using multiplexing) must be initialized. Recall that the 3pi Line Sensors are read as analog devices using a Pololu provided library routine. In order for the measurements to be valid, another library routine must be called to calibrate the Line Sensor readings. A routine that truly calibrates the line sensors themselves will be discussed in Chapter 5. Until that routine is called, our code will assume that the line sensors are disabled based on the value of the variable `LineSensorsEnabled`. In this initialization routine, our goal is to set up the basic variables so that the IR and bumpers can be read.

Basically, the initialization code displayed in Figure 4.12 calls a Pololu provided initialization routine that tries to establish maximum and minimum values for each line sensor with the robot's underbelly lights both on and off. As mentioned, we are not trying to truly find these min/max values here, but calling the routine creates the memory locations to store the data. Pointers to these memory locations are created, and then used to set the min/max values for the lights-off condition to 0 and 1023 respectively. This allows the IR and bumpers to be read properly since they are both read with the underbelly lights off (effectively disabling the line sensors, which need the lights on).

While this code allows the IR and bumper sensors to be read, it does provide the initialization necessary to read the line sensors. The code to handle that will be invoked with a special command and explained in detail in Chapter 5. The line sensors cannot be used until that happens, so the variable LineSensorsEnabled is set to zero when the 3pi is first turned on. You might be wondering why we don't always initialize the line sensors. The full details will be given in Chapter 5, but let's examine this situation briefly now.

In order to initialize the line sensors, the robot must be placed on a white surface over a dark line so that it can determine the actual values read when a line is encountered. Unless your application utilizes a line, it makes no sense to make the user wait for line sensor calibration *every* time the robot is turned on. Therefore, the robot will normally only enable the IR and bumper sensors, and assume that the RobotBASIC application will request additional initialization if it is needed. This may sound complicated, but it is actually very easy and intuitive once you see it in action.

Finally, the initialization routine plays a *welcome* tune on the 3pi's speaker, makes sure the motors are all off, and sets the 3pi's communication baud rate to 9600. Experimentation proved that setting the baud rate higher has a negligible effect on the overall data rate between RobotBASIC and the 3pi, due to the Bluetooth delays associated with switching from transmit to receive mode. Even at 9600 baud, the time taken to send the data is minimal compared to the Bluetooth delays.

```
void initialize()
{
// This must be called at the beginning of 3pi code, to set up the
// sensors. Pololu uses a value of 2000 for the timeout, which
// corresponds to 2000*0.4 us = 0.8 ms on our 20 MHz processor.
   pololu_3pi_init(2000);

// set up Direction Register when necessary for new pin assignments
   DDRB |= (1<<Servo);      // set pin to output for servo control
   DDRB &= ~(1<<Beacon);    // set pin to input to read beacon data
   DDRD |= 1<<MuxCtl;       // set pin to output to control MUX circuits

// initially set Servo and Mux outputs LOW
   PORTB &= ~(1<Servo);
   PORTD &= ~(1<MuxCtl);

// Compass is initialized here... details omitted (given in Chapter 6)

// Note: The analog pins for reading the battery voltage and Distance value
//       are already setup properly, as is PC5 which controls the infrared
//       source for the line sensors (which must be turned OFF when reading
//       the bumpers or the infrared parimeter sensors.  PD0 and PD1 are also
//       properly set up for serial communication by the standard 3pi
//       initialization

// do a simple sensor calibration... only enables IR and BUMPERS
   PORTB |= (1<<4);  // low to disable extended sensors
   calibrate_line_sensors(IR_EMITTERS_ON_AND_OFF);
   delay_ms(2000);
// create pointers to memory space used to store min/max data
   unsigned int* MINON = get_line_sensors_calibrated_minimum_on();
   unsigned int* MINOFF = get_line_sensors_calibrated_minimum_off();
   unsigned int* MAXON = get_line_sensors_calibrated_maximum_on();
   unsigned int* MAXOFF = get_line_sensors_calibrated_minimum_off();
// Make OFF calibration full scale
   for (int j=0;j<5;j++)
      {
      *(MINOFF+j)=0;    //*(MINON+j);
      *(MAXOFF+j)=1023; //*(MAXON+j);
      }

// Play welcome music
   play_from_program_space(welcome);
   delay_ms(1000);

   set_motors(0,0); // make sure motors are initially off
   serial_set_baud_rate(9600);
   delay_ms(200);
}
```

**Figure 4.12**: This code provides the initialization needed for the 3pi program.

# 4.16 Summary

In this chapter you have learned:

- ❑ How to set up I/O pins on the 3pi to function as either an input or output pin.
- ❑ How to set and clear the value of an I/O pin.
- ❑ How the 3pi code receives and decodes commands from RobotBASIC.

❏ How each of the standard commands for the external robot are implemented.

❏ How data for the time-sensitive sensors is collected and passed back to RobotBASIC.

❏ What initialization is required by the external robot, and how it is accomplished.

# Chapter 5

# Secondary 3pi Commands

The previous chapter explained how the 3pi robot communicates with RobotBASIC, and how primary commands are identified and carried out. While these commands provide most of the primary functionality required, it would be nice to have extended capabilities that could give programmers additional control.

## 5.1 The rCommand Function

In order to make it easy to implement such features, RobotBASIC provides the function rCommand which can be used as shown in the example below.

```
x = rCommand(opcode,data)
```

Recall from Chapter 4 that RobotBASIC's built-in protocol causes two bytes (a code indicating the desired operation and a data parameter associated with that operation) to be sent out over the active serial port. This simply meant that simulator statements such as rForward and rFeel can be configured to work with either the simulator or a real robot. Recall also, that a real robot should respond to these commands by sending back five appropriate bytes of data.

The rCommand function allows you to create your own commands to handle situations specific to your particular robot. To use `rCommand`, you must specify two parameters to be sent to the robot (an operation code and its associated data) and the robot will be expected to return five bytes just as it does with the standard instructions. In the above example, these five bytes will be placed in a string and stored in the variable `x`.

Since the RobotBASIC protocol automatically sends nearly all simulator commands to an activated robot, you might be thinking that there are only a handful of situations where `rCommand` would be helpful. This is not the case. This chapter will provide the code for numerous extended commands for the 3pi. Future chapters will show the details of how these commands can be used in RobotBASIC-3pi applications.

Each of the extended commands will have a `case`-block inside the same `switch` structure described in Chapter 4.

## 5.2 Extended Commands

We implemented 17 extended commands that make sense for the 3pi robot. These commands were assigned the opcodes from 110 though 126, although any unused opcodes could have been used. Each of these commands will be discussed individually in the sections below. You may wish to add new commands of your own or to modify these commands to deal with certain situation differently than we did. Ideally, there is enough variety to these commands to demonstrate enough principles so that you can easily create the system you want for your robot.

## 5.3 Playing Sounds and Tunes on the 3pi

The 3pi software provides the capability to play a variety of sounds, notes and even short songs through the robot's speaker. If we give your RobotBASIC programs the ability to utilize this capability, then the 3pi can be more interactive.

For example, you might want the robot to play a short tone or beep when it finds a beacon or when the battery voltage drops below a preset level. We assigned 111 as the opcode for this function. Since there is only one byte of data, we cannot specify parameters such as frequency and duration. Instead, the data parameter is used to specify which sound to play. The code in Figure 5.1 can help clarify this approach.

```
case 111:
  if (argument==1) play_from_program_space(welcome);
  if (argument==2) play_from_program_space(go);
  if (argument==3) play("c");
  if (argument==4) play("cdefedc");
break;
```

**Figure 5.1**: This code plays selected tones and tunes.

Since the argument data is 8 bits, you can define up to 256 different sounds to be played. The 3pi library functions provide the ability to play notes or sequences of notes directly from a standard string or from a string stored in the program memory (thus saving RAM).

The code in Figure 5.1 shows examples of how various sounds can be played. The Pololu documentation provides details on how to utilize numerous features including sharps and flats, half and full notes, etc. Study it carefully and you should be able to create nearly any sounds you might want.

## 5.4 Calibrating the Servomotor

Recall that the distance-measuring sensor is mounted on a servomotor, allowing it to scan in any direction in front of the robot. The default position of the motor should face the sensor directly forward. This position is selected when the robot is initialized by sending a series of 1250 microsecond pulses to the servomotor (refer to Figure 4.2).

Due to minor variations between servomotors, different motors might point in slightly different directions when given the same control pulses. In order to compensate for this, the value of the variable `ServoOffset` is always used to adjust the width of the pulse being sent to the servomotor. This allows a RobotBASIC application to add or subtract up to 127 microseconds from the default centered time period, thus ensuring that the default position of the servomotor is directly forward.

We assigned 112 as the opcode for an extended command that could establish a value for `ServoOffset` as shown in Figure 5.2. The code simply copies the argument data into `ServoOffset`. Since the argument is always passed from RobotBASIC as a positive number, it is adjusted to allow it to act as a signed parameter.

```
case 112:
  ServoOffset=argument;
  if (ServoOffset>127)
    ServoOffset-=256;
break;
```

**Figure 5.2**: This code ensures that the default position for the servomotor is correctly centered.

## 5.5 Calibrating the Line Sensors

Recall that the 3pi Line Sensors are read as analog devices using a Pololu provided library routine. In order for the measurements to be valid, another library routine must be called to calibrate the Line Sensor readings. Basically, the initialization routine tries to determine what the maximum and minimum values are for each sensor, for the current conditions (the color, or shade of gray, of lines to be followed as well as the floor color itself can alter these max/min values).

During normal startup, our 3pi software assumes max/min values of 1023 and 0. This allows the multiplexed sensors (IR and bumper) to work properly, but the Line Sensors themselves will not operate reliably without establishing true max/min values. In order to do this, we need to read each line sensor while it is over both a dark line and a light floor area (an area without a line).

We could do this by requiring someone to physically move the 3pi so its sensors pass over a line, but this would be much too cumbersome. Instead, we will expect that the human operator should place the robot on a single line with plenty of white space on either side. Our software will then use the robot's motors to turn it left and right so that all sensors are moved over the line. During this movement, the sensors will be read and the maximum and minimum values will be recorded and then used to adjust future readings thus ensuring their reliability. Remember, when the 3pi is first initialized, the variable `LineSensorsEnabled` prevents the Line Sensors from even being used. Once the calibration routine is called, the value of the `LineSensorsEnabled` will be adjusted to allow operation of the Line Sensors.

Figure 5.3 shows the calibration code, which is executed when the opcode is 113. At the beginning of this routine a signal is sent to the multiplexing circuitry to ensure that the bumper and line sensors are both disabled. Next, we make sure the motors are turned off before the actual calibration begins.

An `if`-statement inside the first `for`-loop forces the robot to turn both left and right for approximately 90° in each direction, ensuring that all five line sensors pass over the line (assuming the robot has been set to straddle a line before calibration is begun).

While the robot is turning, the Pololu routine `calibrate_line_sensors()` is used to take readings with the lower IR lights both on and off. For our purposes, we only need the readings with the lights on, but asking for the readings with the lights off too, forces the library routine to create memory space for them (and we do need that memory space).

When the `for`-loop is finished, the motors are turned off, and Pololu routines are used to obtain pointers to the memory space where the maximum and minimum values are stored. Notice that the pointers for the lights-on values are also retrieved in case you want them, but we only need the lights-off pointers. A second `for`-loop uses these pointers to set the max/min values to 1023/0 (just as is done when the 3pi is initialized). The difference here is that the lights-on max/min values are now correctly set too so that the Line Sensors will work properly.

```
case 113:
    PORTB |= (1<<4);    // line high to disable IR and bumper
    set_motors(0,0);
    delay_ms(100);
    // Auto-calibration: turn right and left while calibrating
    for(counter=0;counter<80;counter++)
        {
        if(counter < 20 || counter >= 60)
            set_motors(speed,-speed);
        else
            set_motors(-speed,speed);
            // This function records a set of readings and keeps
            // track of the minimum and maximum values
            calibrate_line_sensors(IR_EMITTERS_ON_AND_OFF);
            delay_ms(20);
        }
    set_motors(0,0);
    // Make OFF calibration Same as ON calibration
    unsigned int*MINON =  get_line_sensors_calibrated_minimum_on();
    unsigned int*MINOFF = get_line_sensors_calibrated_minimum_off();
    unsigned int*MAXON =  get_line_sensors_calibrated_maximum_on();
    unsigned int*MAXOFF = get_line_sensors_calibrated_minimum_off();
    for (int j=0;j<5;j++)
        {
        *(MINOFF+j)=0;    //*(MINON+j);
        *(MAXOFF+j)=1023; //*(MAXON+j);
        }
    // now that the line sensors are calibrated
    // we can enable them by setting the variable to true
    LineSensorsEnabled = 1;
    break;
```

**Figure 5.3**: This code calibrates the light sensors by establishing the max/min values needed.

## 5.6 Calibrate Rotation Times

Recall from Chapter 4 (see Figure 4.4) that the variable `RotTime` is used to make sure the robot's open-loop turns are reasonably calibrated to the number of degrees specified. The default value of 12 should work on most 3pi robots, but we can use opcode 114 to establish a new value for `RotTime`, letting you fine tune the operation of your hardware. This is easily done as shown in Figure 5.4.

```
case 114:
   RotTime=argument;
break;
```

**Figure 5.4**: Rotational turns can be calibrated by setting a new value for `RotTime`.

## 5.7 Calibrate Forward and Reverse Movements

Linear movements of the robot are equated to the pixel distances specified in the simulator commands using the value of the variable `MoveTime` (refer back to Figure 4.3). Opcode 115 is used to create an extended command, as shown in Figure 5.5, for changing the value of `MoveTime`, letting you calibrate forward and reverse movements on your robot. The default value for `MoveTime` is 24.

```
case 115:
   MoveTime=argument;
break;
```

**Figure 5.5**: Linear movements can be calibrated by setting a new value for `MoveTime`.

## 5.8 Calibrating Drift

When the 3pi moves forward or backward, it will only move in a straight line if both wheels are turning at the same speed. The variables `LeftOffset` and `RightOffset` (refer back to Figure 4.3) are used to calibrate such movements to minimize drift. Opcode 116 provides a command for setting new values for these variables, as shown in Figure 5.6.

The code in Figure 5.6 assumes that the 8-bit parameter passed to the 3pi is composed of two parts. The upper 4 bits contains a signed offset for the right wheel while the lower 4 bits contains the offset for the left wheel. The code in Figure 5.5 extracts these offsets and places them in the correct variables.

```
case 116:
  Roffset = argument&15;
  Loffset = argument>>4;
  if (Roffset>7) Roffset=16-Roffset;
  if (Loffset>7) Loffset=16-Loffset;
break;
```

**Figure 5.6**: This code extracts signed offsets for the left and right wheels.

Recall, that if the compass-mode is enabled by the variable CompassEnabled, that the robot's forward movement may also be corrected to keep the robot on its original heading. In environments where stray magnetic field or metal objects might interfere with the compass's operation, the robot might perform better with the compass disabled. If you are not using the compass, the drift compensation described above is critical. Even if you are using the compass, having the wheels moving at nearly the same speed minimizes the need for additional correction.

## 5.9 Calibrating the Compass

We developed special routines for dealing with the compass. They will be covered in Chapter 6. Because of these routines, it is easy to create an extended command to calibrate the compass, as shown in Figure 5.7. The Calibrate_Compass() routine rotates the 3pi while performing the internal calibrations. The rotational speed is based on the current value of **speed**, so it must be passed as a parameter.

```
case 117:
  Calibrate_Compass((int)speed);
  CompassEnabled = 1;
break;
```

**Figure 5.7**: The compass routine used here will be discussed in detail in Chapter 6. Notice the compass is enabled when it is calibrated.

## 5.10 Modifying the Turn Style

The RobotBASIC simulated robot always turns by rotating around its center (turning one wheel forward and the other backward, simultaneously). This movement has the advantage of always being able to turn without bumping into an object. This movement also works with a real robot too, but there are situations where non-rotational turns can be advantageous, especially when the physical robot is controlled through a wireless link with speed limitations.

When following a line, for example, rotational turns coupled with the speed limitations can make the robot move in a jerky manner as it rotates from side to side. An easy solution to this problem is to force the robot to make turns by activating one motor at normal speed and the other motor at some percentage of the first. The more difference in the speeds of the two motors, the greater the turning speed. Later chapters will explore this movement and how it can be used to smooth line-following activities. Figure 5.8 shows the code for implementing an extended command for controlling how turns are to be handled. Refer back to Section 4.7 (Chapter 4) to see how the variable TurnStyle is used to influence the robot's behavior.

```
case 118:
  TurnStyle=argument;
  if TurnStyle>8 TurnStyle=8;
break;
```

**Figure 5.8**: This extended command controls how the real robot executes turns.

## 5.11 Full Stop

While it should not be necessary to issue a Full Stop command to the external robot, having such a choice seemed like an appropriate option. Figure 5.9 shows how this command stops both motors.

```
case 119:
    set_motors(0,0);
break;
```

**Figure 5.9**: This extended command turns off both motors.

## 5.12 Enabling the Compass

Previous Chapters have discussed the idea that the compass may or may not be used to augment both linear travel and rotational turns. Figure 5.10 shows how opcode 121 allows the application program to specify whether the compass should be used or not by establishing a new value for the variable CompassEnabled.

```
case 121:
    CompassEnabled=argument;
break;
```

**Figure 5.10**: This command allows the user to determine if the compass is used during significant movements.

## 5.13 Sensor High-Level

When the Pololu routines are used to read the status of the Line Sensors, IR Sensors, or the Bumpers, the data returned will be between 1023 and 0 depending on the status of the sensor. For example, when the Line Sensors are over a dark line, they should return a value near 1023, while sensors over a white (reflecting) surface should provide a near zero reading.

Readings above some specific level should be considered a logical one and those below that level should be considered a zero. We have experimentally chosen 200 to be the threshold level between a logical one and zero, but there could be situations where another level would be better. The extended command shown in Figure 5.11 allows the level to be altered by setting it to twice the parameter passed (overcoming the one byte limit of 255).

```
case 122:
   HiLevel=2*argument;
break;
```

**Figure 5.11**: This command allows the user to select the dividing point (for Line, IR, and Bumper sensors) between a logical one and zero.

## 5.14 Setting the Distance Mode

Recall from the last Chapter (refer back to Section 4.10) that readings from the Distance Sensor can be reported in inches or pixels, or even in a raw form that allows for maximum resolution. Figure 5.12 shows an extended command that allows you to choose how the Distance Sensor reports its data.

```
case 123:
   DistMode=argument;
break;
```

**Figure 5.12**: This command allows the user to specify how the distance measuring sensor returns its data.

## 5.15 Find a Beacon

RobotBASIC's rBeacon command allows you to write RobotBASIC code that turns the robot until a beacon is found. While this is a workable approach, it could present some speed problems. Let's see why.

If we look for the beacon by rotating the robot using an rTurn command, we cannot turn it again until we know if a beacon has been spotted, and that means waiting for parameters to be returned over the wireless link. One solution to this problem is to create code in the 3pi so that it can be requested to search for a beacon on its own. Since the 3pi code can respond almost immediately to changes in sensory data, the robot can be rotated at a much faster rate without overshooting the beacon.

Many of the extended commands we have discussed have been very simple because they are only establishing a new value for a variable (which in turn controls actions elsewhere in the 3pi code). In this case, the code required (shown in Figure 5.13) is larger because much more has to be done.

When this code executes, the variable `argument` specifies which direction to rotate the robot when looking for the beacon (a value of 1 creates a clockwise turn, any other value produces a counter-clockwise turn). An `if`-statement turns on the motors in the proper direction at the standard speed.

```
case 125:
  // argument specifies direction
  temp1=0;
  if (argument==1)
    {
    set_motors(50,-50);// overcome friction etc
    delay_ms(10);
    set_motors( (speed*2)/3,-( (speed*2)/3) ); // turn right
    }
  else
    {
    set_motors(-50,50);// overcome friction etc
    delay_ms(10);
    set_motors(-((speed*2)/3),(speed*2)/3); // turn right
    }
  while(!(PINB&(1<<Beacon)))  // move away from current beacon
    ;
  delay_ms(15);
  for(temp=0;temp<7000;temp++) // enough time for a FULL revolution
    {
    delay_ms(1);
    if(!(PINB&(1<<Beacon)))
      {
      temp1=1;
      break;  // look for new beacon
      }
    }
  set_motors(0,0);
  SensorBuffer[4]=temp1;
```

**Figure 5.13**: This code rotates the robot until it sees a beacon.

Next, a `while`-loop waits until no beacons are seen (effectively moving away from the current beacon, if it exists). This is important because an application could continue to request this code to move to the next beacon, until the desired one is found. Later chapters will explore this concept in more detail.

A `for`-loop rotates the robot up to 360° while an `if`-statement monitors the state of the beacon sensor and changes the value of a temporary variable to 1 if *any* beacon is detected. At the end of the loop, or if a beacon has been detected, the rotation is stopped and the `SensorBuffer` is updated to indicate if a beacon has been detected.

## 5.16 Turn Robot to a Specific Heading

An application program might want to turn the robot to a desired heading, such as due East. If this is done with standard RobotBASIC commands, the same speed problems discussed in Section 5.13 apply. Because of this, we will provide an extended command that allows the 3pi to move to a desired heading without external intervention. Figure 5.14 shows the code for this command.

```
case 125:
  MoveToAngle(argument*2,speed);
break;
```

**Figure 5.14**: This extended command calls the code in Figure 5.14 to move the robot to a desired heading.

The extended command shown in Figure 5.14 has a limitation that needs further explanation. The desired heading can be anywhere from 0 to 359 degrees, but the data parameter passed to the 3pi is only 8 bits, thus limiting the value passed to 255. Because of this, the RobotBASIC application will be expected to pass the desired heading divided by 2, thus limiting the value to 8 bits. As you can see from Figure 5.14, the data received will be multiplied by 2 before being passed to a function that will perform the actual movement. Although the accuracy of this command is limited, in practice it is not a problem, especially when you realize that motor overshoot does not allow extremely accurate positioning anyway. The code in Figure 5.14 is simple because all the work is done by the subroutine shown in Figure 5:15, which actually moves the robot to the desired heading.

The code in figure 5.15 may not operate as you might expect. A typical algorithm for this situation might involve complicated PID control. Fortunately, a simpler approach proved surprisingly workable.

A for-loop moves the robot towards the specified heading five times, decreasing the motors speed on each pass (thus handling problems with overshooting the destination). Each successive attempt improving on the desired accuracy.

```
void MoveToAngle (int angle, int speed)
{
int i,j,temp;
if(angle<0) angle=angle+360;
for (i=0;i<=4;i++)
  {
  temp=angle;
  if (temp<Read_Compass()) temp=temp+360; //make destination larger
  if ((temp-Read_Compass())<180) // decide which direction to turn
    {
    set_motors(50,-50);// overcome friction etc
    delay_ms(5);
    set_motors(speed-(i*4)-6,-(speed-(i*4)-6)); // turn right
    }
  else
    {
    set_motors(-50,50);// overcome friction etc
    delay_ms(5);
    set_motors(-(speed-(i*4)-6),speed-(i*4)-6); // turn left
    }
  if(i==0)
    while ((Read_Compass()/32)!=(angle/32));
  if(i==1)
    while ((Read_Compass()/16)!=(angle/16));
  if(i==2)
    while ((Read_Compass()/8)!=(angle/8));
  if(i==3)
    while ((Read_Compass()/4)!=(angle/4));
  if(i==4)
    for(j=0;j<6;j++)
      {
      if ((Read_Compass()/2)==(angle/2)) break;
      };
  set_motors(0,0);
  delay_ms(50);
  }
}
```

**Figure 5.15:** This routine moves the robot to a specified heading.

The first if-statement in the loop makes sure that the destination angle is always larger than the robot's current heading. This is necessary for the formula used in the second `if`-statement, which determines if the robot should move clockwise or counterclockwise. Notice that the speed on each pass through the `for`-loop is decreased by a factor of the variable `i`. Furthermore, as the speed is decreased the code tries to move the robot to a more accurate position on each pass using `while`-loops.

On the final pass through the loop, the robot is turning very slowly; so slowly in fact, that the robot might possibly stall. The for-loop at this point in the code prevents the code from hanging by only attempting the move six times. The `break` statement allows the `for`-loop to be exited early if the correct heading has been reached.

## 5.17 Setting the Robot's Speed

The simulator's speed is established using the `rSpeed` command. Section 4.14 (Chapter 4) showed how to ensure that `rSpeed` statements have no effect on the remote robot - after all, just because you might want to slow down or speed up the simulation during algorithm development does not mean you want to do the same thing to the robot.

There are times though, that you might want to increase or decrease the speed of the 3pi itself. Opcode 127, as shown in Figure 5.16, provides this capability. The default speed is 40.

```
case 126:
  speed = argument;
break;
```

**Figure 5.16**: This code allows the 3pi's speed to be increased or decreased as desired.

## 5.18 Setting the Timeout

Normally the 3pi has a timeout of 500ms. If the 3pi does not receive a command within this time period, it stops the robot's motors and waits again for another command. Recall that normally, once the motors are started, the 3pi keeps them running until specifically told to stop them (to prevent a jerky motion). This means that if a RobotBASIC program were to hang, the robot would continue to move. The timeout period causes the robot to always stop if it does not receive commands. The code in Figure 5.17 allows the user to change the timeout period.

```
case 110:
  MyTimeOut = argument;
break;
```

**Figure 5.17**: This allows the user to set the 3pi's Timeout period.

## 5.19 Reset Defaults
Many of the extended commands discussed in this chapter establish new values for variables that control how the robot operates. The command shown in Figure 5.18 resets all these parameters to their original values.

```
case 120:
  speed=40;
  RotTime=12;
  MoveTime=24;
  Loffset=0;
  Roffset=0;
  TurnStyle=0;
  ServoOffset=0;
  DistMode=0;
  HiLevel=200;
  CompassEnabled=0;
  LineSensorsEnabled=0;
  MyTimeOut=500;
break;
```

**Figure 5.18**: This code resets all control variables to their default values.

## 5.20 Summary
In this chapter you have learned:
- ❑ How rCommands can be used to change the 3pi's operational modes.
- ❑ The details of how 17 different rCommands were implemented.

# Chapter 6

# The Compass

The code in previous chapters called special compass routines when the compass's functionality was needed. Now it is time to look into the details of those routines.

## 6.1 Interfacing the Compass

In general, external devices can be interfaced to a main processor in two ways, using parallel or serial ports. Parallel ports have the advantage of being extremely fast because multiple bits (typically 8) can be transferred simultaneously between two devices. Serial ports, on the other hand, transfer the bits one at a time, over the same wire or other transmission media. Serial ports are often easier and less expensive to implement.

Serial ports of the past, such as the RS-232 standard, used asynchronous transfers. This simply means that it was up to each individual device to create a clock pulse of the proper timing requirements. Today's newer technology often uses synchronous transfers where a *master* device is responsible for the clock pulses controlling the transfers both to and from a *slave* device. One of these new synchronous protocols is called $I^2C$. The synchronous nature of this protocol makes it potentially far faster than asynchronous technology.

## 6.2 The I²C Protocol

I²C allows a single master to control transfers between itself and one or more slave devices as shown in Figure 6.1. Only two wires (plus ground) are required.

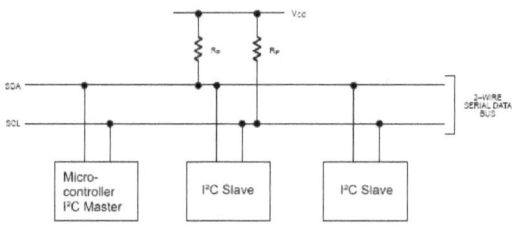

**Figure 6.1**:Transfers using the I²C protocol require only two wires.

The fast speed of this interface, plus needing only two wires (and thus only two I/O pins), makes it an ideal solution for connecting peripheral devices to micro-controllers such as the one used on the 3pi.

One of the two wires discussed above carries the clock pulses generated by the master. The other line carries the data, which has several implications. Since only one wire is used for data, it is the responsibility of each device to make the port pin connected to this wire either an input-pin or an output-pin, depending on what is needed. The problem is that just switching a pin between input and output can be interpreted as a change in data (high-to-low or low-to-high). For that reason a unique arrangement is often used to handle this situation.

Notice the two pull-up resistors in Figure 6.1. They connect each of the lines (one clock and one data) to the 5V power supply. This effectively makes each of these lines float to a high (or logical one) condition unless something pulls that line low. If an output port connected to this line is set to zero (low) then we can pulse this line high or low by simply making the port pin an input (which lets the line float high) or an output (which pulls the line low).

This lets the master pulse the line as necessary. For example, the port pin could be changed between input and output in order to place the correct data (to be sent to a slave) on the data line. More importantly though, when the port pin is in the input state, the master can read data placed on the line by the slave.

In order to ensure compatibility of $I^2C$ communications, strict standards have been established that control the details of all transfers. These include such things as *start* and *stop* conditions as well as the ability to specify addresses to indicate which slave is being used. An Internet search can provide you with such information, should you want to understand the low-level details. Fortunately, though, we have created 3pi routines that can handle $I^2C$ communications for you so that you do not have to understand all the details.

Some readers may be disappointed by the above statement, because they would like to see every little detail explained. If you feel this way, consider the following argument.

## 6.3 The Power of Modularity

Earlier chapters used many Pololu-provided routines to handle a wide variety of tasks. There were routines, for example, that handled RS-232 communications with the Bluetooth transceiver, routines that read the status of the 3pi's line-sensors, and routines that controlled the speed of the 3pi's motors. Each of these routines gives us great power to accomplish larger and more exciting goals, because we do not have to worry about every minute detail.

Imagine someone using a bulldozer to accomplish the goal of creating the proper landscape for a construction project. While there might be some benefits for the bulldozer operator to understand the basic principles of hydraulics, it should be clear that the operator does not have to understand every detail of the bulldozer's internal operation. This is generally true of most things we use today. Engineers often benefit from test equipment (such as oscilloscope) without considering the details of how the device actually works. Millions of people get their news and entertainment from the Internet or their television without giving any thought to how either of these systems accomplishes its tasks.

The fact that we live in a modular world allows us to do many things our ancestors were not able to do, just as they were able to do more than their ancestors. The reason is tools. Each new generation does not have to start from scratch. Instead, they get to build their world using the tools that were built by the previous generations.

Only a few centuries ago, electricity was an unknown entity. The tools developed by Benjamin Franklin led to an understanding of its nature, and that led to the discoveries of Edison, Hertz, Maxwell, and Tesla, which in turn led to the vacuum tube and then the transistor, which led to the integrated circuit and eventually the microprocessor.

Hobbyists only a few decades in the past required a degree in electronics and significant programming skills if they wanted to build a robot with even moderate intelligence. The tools available today though, allow average people (not just engineers and computer scientists) to experiment with many advanced subjects. The electronic compass used in our Pololu Project is a great example of this situation.

## 6.4 The HMC6352 Compass

Ten years ago, if you wanted an electronic compass you would have to design and construct one using hall-effect devices. Ten years before that, you would have had to construct your own hall-effect sensors. Today all of that technology is available in the HMC6352 compass from Honeywell. While the HMC6352 is a fantastic tool that provides magnetic headings with resolutions less than one degree, it possesses attributes that can make it difficult to use for the average hobbyist (for example, the HMC6352 cannot be powered with the industry standard 5V supply). Parallax enhanced the basic Honeywell chip through a carrier board that makes it easier to interface and power with a standard 5V supply, thus making the tool accessible to many more people.

The compass module communicates using an $I^2C$ interface as described above, making using the compass extremely easy for anyone that has either a hardware or software $I^2C$ interface. Think about what this means.

You, the hobbyist, can utilize fantastic technology in your projects without having to understand the details of that technology. You do not have to know how the compass module works internally - you do not even have to understand the complexities of the $I^2C$ interface because (in this case) we have provided that for you.

This is an important characteristic of today's technology. Even if you are not an engineer or master programmer you can utilize many modern tools in your projects. You may have to do a little research to find tools that can help you implement your goals, but companies like Pololu, Parallax, and many others, are constantly making it easier than ever to achieve your goals.

## 6.5 The Software Interface

Figure 6.2 shows the basic, fundamental tools needed to pulse the data and clock lines. We have tried to make these tools (and those that follow) as generic as possible so that they can be used with other $I^2C$ sensors should the need arise. Some of the code might be confusing, especially if you are not an engineer that has studied the specs of the $I^2C$ interface, but remember the premise of this chapter - you do *not* need to understand every detail in order to utilize this technology.

```
void InitI2C(void)
   {
   // Compass I/O setup prepares both ports must float
   // Initial data is set LOW so any time the PIN is an OUTPUT
   //     it will be LOW
   // When a PIN floats, EXTERNAL pull-up's allow it to appear HIGH
   DDRD |= CompassClock;          // make both pins output
   DDRD |= CompassData;           // so we can set the DATA register
   PORTD &= ~(1<<CompassClock);   // Set both outputs LOW
   PORTD &= ~(1<<CompassData);
   DDRD &=  ~(1<<CompassClock);   // Set both to INPUT so they FLOAT
   DDRD &=  ~(1<<CompassData);
   }

void SetSDA_Input(void)   // sets the data line high
   {
   DDRD &=  ~(1<<SDA_PIN);
   delay_us (500);
   }

void SetSDA_Output(void)   // sets the data line low
   {
   DDRD |=  (1<<SDA_PIN);
   delay_us (500);
   }

void SetSCL_Input(void)   // sets the clock line high
  {
   DDRD &=  ~(1<<SCL_PIN);
   delay_us (500);
  }

void SetSCL_Output(void)  // sets the clock line low
  {
   DDRD |=  (1<<SCL_PIN);
   delay_us (500);
  }
```

**Figure 6.2**: This code represents the tools needed
to implement the I$^2$C interface.

Figure 6.3 shows how the tools of Figure 6.2 can be used to build new tools that produce fundamental actions (start, stop, ack, and nack) required by the I$^2$C specifications.

Figure 6.4 shows how the low-level tools can be used to implement routines that can read and write bytes on the I$^2$C bus. These routines can then be used to create the user-level tools for reading and calibrating the compass as shown in Figure 6.5.

Defined constants make the compass code more readable. They are shown in Figure 6.6. Some constants are not used here, but might be useful to you, should you need to interface other I$^2$C devices.

```
void I2C_Start(void)
  {
  //I2C start condition.  SDA transitions high to low while the clock is high.
  //Assumes SDA and SCL are both floating high
  SetSDA_Output();  //Pull SDA low
  SetSCL_Output();  //Pull SCL low
  }

void I2C_Stop(void)
  {
  // I2C stop condition.  SDA transitions low to high while the clock is high.
  // Assume
  SetSDA_Output();        //Pull SDA low
  SetSCL_Input();         //Let SCL float high
  SetSDA_Input();         //Let SDA float high
  }

void I2C_ACK(void)
  {
  //Acknowledge for read; use between reading successive bytes
  SetSDA_Output();        //Pull SDA low
  SetSCL_Input();         //Clock out an acknowledge bit
  SetSCL_Output();        //
  SetSDA_Input();         //Let SDA float high
  }

void I2C_NACK(void)
  {
  // Non-acknowledge for read; use after reading last byte
  SetSDA_Input();         //let SDA float high
  SetSCL_Input();         //Clock out an acknowledge bit
  SetSCL_Output();        //
  }
```

**Figure 6.3**: Mid-level tools are produced  from those shown in Figure 6.2.

```
int I2C_Read(void)
  {
  //Read a byte from the I2C device, I2C_ReadByte will hold the read byte
  int I2C_ReadByte = 0;
  int i;
  SetSDA_Input();          //Let SDA float
  for( i=0; i<8;i++)       //for 8 bits starting with the MSBit
  {
    I2C_ReadByte = I2C_ReadByte << 1;  //shift last read bit to left
    SetSCL_Input();                    //let SCL float
    if(PIND &(1<<SDA_PIN))             //read port and mask for SDA pin
      I2C_ReadByte++;                  //if high then increment to set bit
    SetSCL_Output();                   //Pull SCL low
  }
  return I2C_ReadByte;
}

int I2C_Write(int I2C_WriteByte)
  {
  //Write a byte to the I2C device I2C_WriteByte is the byte
  //to write and I2C_AckBit will hold the Ack from slave
  int i;
  for( i=7;i>=0;i--)  //for 8 bits starting at the MSB
    {
    if (I2C_WriteByte&(1<<i))    //examine bit in the Data byte
      SetSDA_Input();            //if high then set SDA to input(high)
    else
      SetSDA_Output();           //if low then make SDA output and thus low
    SetSCL_Input();              //Pulse SCL high then low
    SetSCL_Output();
    }
  SetSDA_Input();                //Let SDA float
  SetSCL_Input();                //Start a pulse on SCL
  i = 0;
  if(PIND & (1<< SDA_PIN))       // read the SDL pin for the ack status
    i = 1;                       //set i to 1 if SDL is high
  else
    i = 0;
  SetSCL_Output();               //Finish a pulse on SCL
  return i;
}
```

**Figure 6.4:** These tools read and write byte data for the compass.

```
int Read_Compass(void)
  { // returns compass Heading
  int Heading;
  InitI2C();
  I2C_Start();
  I2C_Write(HMC6352_WRITE);
  I2C_Write(GET_HEADING);
  I2C_Stop();

  I2C_Start();
  I2C_Write(HMC6352_READ);

  delay_ms(12); //GET_HEADING_DELAY);
  Heading = I2C_Read() << 8;
  I2C_ACK();
  Heading = Heading |I2C_Read();
  I2C_NACK();
  I2C_Stop();
  return Heading/10;
  }

void Start_Calibration(void) // starts compass firmware
  {
  I2C_Start();
  I2C_Write(HMC6352_WRITE);
  I2C_Write(START_CALIB);
  I2C_Stop();
  delay_us(START_CALIB_DELAY);
  }

void Finish_Calibration(void)
  {
  I2C_Start();
  I2C_Write(HMC6352_WRITE);
  I2C_Write(END_CALIB);
  I2C_Stop();
  delay_us(END_CALIB_DELAY);
  }

void Calibrate_Compass(int speed)
  {
  set_motors(0,0); // make sure motors are off
  delay_ms(300);
  Start_Calibration();
  // rotate 3pi during the calibration proceedure
  set_motors((-speed*3)/4,+(speed*3)/4);
  delay_ms(20000);
  set_motors(0,0);
  Finish_Calibration();
  }
```

**Figure 6.5:** These routines can read and calibrate the compass.

```
#define   SDA_PIN            PORTD2   //compass data pin
#define   SCL_PIN            PORTD4   //compas clockin  pin
#define   HMC6352_READ       0x43     //HMC6352 read address
#define   HMC6352_WRITE      0x42     //HMC6352 write address
#define   START_CALIB        0x43     //Start calibration routine
#define   END_CALIB          0x45     //Finish calibration routine
#define   GET_HEADING        0x41     //Read calculated heading
#define   GET_HEADING_DELAY  6        //6   ms
#define   START_CALIB_DELAY  10       //10  µs
#define   END_CALIB_DELAY    1400     //14  ms
#define   ENTER_SLEEP        0x53     //Enter sleep mode
#define   WAKE_UP            0x57     //Exit sleep mode
#define   MODE_ADDRESS       0x4E     //RAM/EEPROM address
#define   UPDATE             0x4F     //Update mode
#define   SAVE_MODE          0x4C     //Write the operational mode
#define   WRITE_EEPROM       0x77     //Read a byte from EEPROM
#define   READ_EEPROM        0x72     //Write a byte to EEPROM
#define   WRITE_RAM          0x47     //Read a byte from RAM
#define   READ_RAM           0x67     //Write a byte to RAM
#define   READ_WRITE_DELAY   70       //70  µs
#define   ENTER_SLEEP_DELAY  10       //10  µs
#define   WAKE_UP_DELAY      80       //80  µs
#define   UPDATE_DELAY       6000     //6   ms
#define   SAVE_MODE_DELAY    125      //125 µs
```

**Figure 6.6:** Defined constants can make programs more readable.

The `Calibrate_Compass` routine in Figure 6.5 deserves additional explanation. It calls other routines to force the compass to start and stop its internal calibration. In the meantime, the 3pi's motors are turned on to slowly rotate the robot during the calibration process (as required by specification sheet of the compass). The process takes about 20 seconds, so we must extend the timeout period used by RobotBASIC when this command is to be used. This will be discussed in Chapter 9.

## 6.6 Summary
In this chapter you have learned:
- About the I$^2$C interface and how it differs from other serial protocols.
- How the I$^2$C interface can be used to interface the HMC6352 compass to our 3pi robot.
- How modular techniques allow hobbyists to utilize technologies despite not fully understanding their internal operation.

# Chapter 7

# Introductory Control of the 3pi from RobotBASIC

A t this point in the book, we have explored the physical and electrical modifications of the standard 3pi robot, as well as the 3pi firmware that allows the robot to communicate with RobotBASIC. This combination of hardware and software creates an ideal platform for developing robotic algorithms. Later chapters will examine more complex programming examples, but first we need to explore some introductory topics.

## 7.1 Basic Control

If you were introducing someone to the RobotBASIC simulated robot, you might use a program such as the one shown in Figure 7.1.

> **Note:** This book assumes the reader has at least minimal experience programming with RobotBASIC. If you do not, and have trouble understanding any of the examples, we suggest you explore our HELP file, the tutorials on our web page, and/or one of our introductory books.

```
gosub InitializeBackground
rLocate 400,300
rInvisible Green,Red
rPen Down
rSpeed 10
for i=1 to 4
  rForward 100
  rTurn 90
next
end

InitializeBackground:
  SetColor Red
  call DrawCircle(400,300)
  call DrawCircle(400,200)
  call DrawCircle(500,200)
  call DrawCircle(500,300)
  Line 400,300,400,200
  LineTo 500,200
  LineTo 500,300
  LineTo 400,300
return

sub DrawCircle(x,y)
  r=5  // radius of circle
  circle x-r,y-r,x+r,y+r,Red,Red
return
```

**Figure 7.1**: This simple program demonstrates how to control the simulated robot.

If you run the program in Figure 7.1, you will see four circles composing the corners of a red square. The robot will move around the square, stopping and turning when it reaches the center of each circle. The robot will leave a green trail as it moves, which will overlap the red lines that form the square. This behavior represents how an *ideal* robot should respond to this program. Unfortunately, most robots are *not* ideal.

Real robots have to contend with many factors that affect their behavior including such things as friction and motor constants and efficiencies. These factors prevent a real robot from being able to duplicate its actions. For example, a real robot will usually not move in a straight line because one motor will move slightly faster than the other motor. This could be due to slight differences in friction on some areas of the floor or perhaps slight variations in the motors themselves or even in the circuits used to drive each motor.

This limitation is exactly why real robots *must* use sensors to constantly modify their behavior. If no sensors are used, we refer to the control as *open-loop*. A *closed-loop* system is much better because, after each movement is executed, sensory data is used to influence the next movement. This means the robot movement is constantly in a state of correction.

The program in Figure 7.1 uses open-loop control, in that we tell the robot to move a given distance or to turn a specified amount, and we *assume* it does exactly what it is told. This works very well for our simulated robot, because it generally performs exactly as it is told, but doing so does not help us understand the behavior of real robots. Fortunately, our simulated robot has the option of acting more like a real robot.

If you place the command `rSlip 15` in the program of Figure 7.1 (place the command immediately before the for-loop) then the simulated robot will have a random error (up to 15%) associated with all of its movements. If you add the command and run the program, you will see a response similar to the one shown in Figure 7.2. Every time you run the program you will get a different response, because the error generated is truly random.

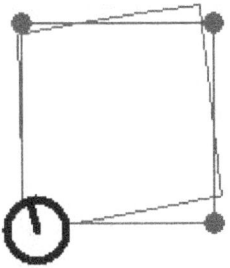

**Figure 7.2**: When an `rSlip` command is executed, the robot's movements are no longer accurate or repeatable.

## 7.2 Controlling the Pololu 3pi

We can use the program shown in Figure 7.1 to control the 3pi robot by inserting an `rCommPort portnumber` command in the program (anywhere prior to the `rLocate` command). For this command to work, you must have a compatible USB Bluetooth dongle installed on your PC and a properly equipped 3pi. After installation, you should see a dialog box similar to the one in Figure 7.3 that provides you with a `portnumber` to insert in the `rCommPort` command (in this case, the full command would be `rCommPort 27`).

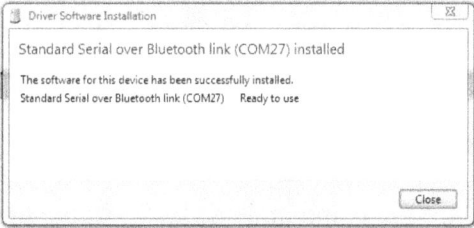

**Figure 7.3**: After installation of a Bluetooth device the dialog box should provide you with a port number.

When you turn on the 3pi robot, a Bluetooth connection should be established automatically, although you might have to manually create the connection the first time. To do so, just turn on the robot, and click the Bluetooth icon in your Window's system tray and utilize the appropriate menu choices to show and select the Bluetooth device you used on your 3pi.

The remainder of this book will examine how programs written for the RobotBASIC simulator can control the 3pi robot. Of course, you do not have to utilize the simulator at all, but why wouldn't you? The simulator is an ideal debugging tool, allowing you to quickly pre-test your programming ideas without running down the batteries or worrying about crashing your physical robot into the wall or down a flight of stairs.

Before we proceed, it is important that you truly understand the difference between open and closed-loop control of our robot.

# 7.3 Open-Loop Control

The term *open-loop* control simply means that the device being controlled does not provide any feedback to the controlling device. A simple example can make this clear. Suppose, for example, we tell the robot to move forward ten inches. Just because we issued the order, we have no way of knowing if it was *properly* carried out.

Even if the robot received the order, and turned the motors on appropriately, the results can still be unpredictable. Perhaps the wheels slip on a smooth floor or maybe some heavy object in the robot's path prevents it from moving at all.

Imagine you have a friend wearing a headset, and you are giving him orders through a microphone from another room. Assume you give the following orders.

- Take ten steps forward
- Turn right
- Take five steps forward
- Do an about-face
- Take five steps forward
- Turn left
- Take ten steps forward
- Do an about-face

If your friend executed all the orders properly he would end up at exactly the same spot he started from, and be facing exactly the same direction as he was in the beginning. But how do you know the orders were executed properly?

Think about it. There are many reasons that your friend might not respond properly. If you issue the orders too fast (maybe you don't wait long enough before telling him to turn right, for example) then he may get behind and not hear some of your orders (and thus not even try to carry some of them out). Also, you are probably assuming when he moves forward that he does so in a straight line and that each step he takes is exactly the same length – both of which are probably not true (especially if he is blindfolded so that he too, is not receiving any feedback).

We can ensure that the friend does not miss any commands by simply waiting till he is done with the last order before we issue another. This is easily accomplished by having him provide feedback by simply having him say "done" or something similar when he finishes a task so we know when to proceed. RobotBASIC's internal communication protocol handles this for us by always waiting for the remote robot to respond before continuing through the program. This simple procedure closes the communication loop between RobotBASIC and the remote robot and helps ensure that the desired actions actually get carried out.

Even with the above action, though, we still have no way of knowing if the robot actually moves the required distance, or if the movements are in a straight line. Once potential solution to such problems is to place incremental encoders on each wheel so that we know the motors have indeed moved properly, but even that won't work if the wheels slip on the work surface. This may seem like an insurmountable problem, but there is a solution.

## 7.4 Sensory Feedback

Assume for a moment, that you wanted to command your friend to follow along a brick wall (until he finds a gate). What kind of orders would you issue? Remember, you cannot just tell him to follow the wall; you must explain how to accomplish what you want done.

One solution would be to have your blindfolded friend use his sense of touch to keep track of where the wall is in relationship to his body. You might, for example, issue the following orders to him. Assume he is currently at the wall, and facing it, and that you want him to follow the wall to his right. Also assume he is giving you appropriate feedback after each order.

```
Turn right
(START FOLLOWING)
   Take one small step forward
   Ask him if he can feel the wall with his left hand
   if NO, then have him turn left slightly
   if YES, then he should turn right slightly
   Ask him if he has gotten to the gate yet
   If NO, then go up to START FOLLOWING
   (and repeat all the steps in this loop)
IF YES, then we have accomplished the goal.
```

Notice, that with these instructions, it does not matter if your friend moves in a straight line. It does not matter if the steps are all the same size or even how much he turns. This is true because he is not just blindly following a set of commands. Instead, the commands he executes are based on feedback provided by the environment itself. This methodology is the only way to create intelligent machines. This idea will be explored more in later chapters.

## 7.5 Calibrating the Robot's Drift

Even though it is vital that we utilize closed-loop control of our robot, it is not unreasonable for us to expect the robot to make a reasonable attempt to carry out our commands.

For example, if we told the robot to move forward 20 units, we might expect it to be off a little, but certainly it should move forward something close to 20 units. It might also drift a little left or right, but again, the movement should be within some reasonable tolerance.

To that end we need some way of calibrating the motor movements. When writing the 3pi firmware, we could have fine-tuned the speed for each motor to make our robot move as properly as possible, but that would mean that the firmware would have to be rewritten for every robot. Recall from Chapter 4, we integrated the following solution into our code.

```
set_motors(speed+Loffset,speed+Roffset);
```

The speed of both motors should be something close to the amount specified by the variable `speed`. But, each motor's actual speed can be adjusted by changing the values of `Loffset` and `Roffset`. Negative numbers for these values will make that wheel a little slower while positive numbers will make it a little faster. Obviously, the offset parameters should be small compared to the value of `speed`.

> **✍ Note:** Normally the 3pi firmware uses the compass to make the robot move in a reasonably straight line. The compass-compensation should be disabled (the default condition) with the `rCommand` in order to see what offset values might be needed. If movements are properly calibrated without a compass, then compass-compensation will be more effective when it is turned on.

Recall also, that we implemented a special code into our 3pi firmware (see Section 5.6 in Chapter 5) to enable us to change the values of these offset variables. We can use RobotBASIC `rCommand` to send a custom code to the remote robot as well as a single 8-bit parameter that, in this case, can alter the value of the offset variables.

The code for this command is 116 and the parameter byte uses the upper and lower nibbles to change the values of the left and right offset variables. It is easier to see the values of the upper and lower nibbles of the data byte if we use 0x to indicate that the number is in a hex format. Lets look at some examples.

```
rCommand (116, 0x10)  // increases speed of left wheel by 1
rCommand (116, 0x20)  // increases speed of left wheel by 2
rCommand (116, 0x02)  // increases speed of right wheel by 2
rCommand (116, 0xF1)  // decreases left by 1, increases right by 1
```

As you can see, the value of each nibble is used to alter the value of the appropriate offset variable. Negative numbers should be represented in two's complement format (F, for example, would indicate negative one). In general, you can probably calibrate your robot's motors by increasing the speed of either the left or right motor by one or two. The valid range of choices is -8 to +7, which should be far more than you need.

It is important to realize the value of this type of calibration. Using rCommands, the user can adjust key parameters inside the 3pi's firmware, to fine-tune the robot's behaviors. This means a school, for example, can have several robots, and each can use exactly the same firmware because the application programs (written in RobotBASIC) can utilize a library of subroutines composed of appropriate rCommands that calibrate the necessary parameters.

This may seem like a complex idea, but it is actually very easy to implement. In fact, rather than try to explain everything here, the details will be demonstrated in future programs throughout this book.

## 7.6 Additional Calibration

If you examine the various rCommand codes discussed in Chapter 5, you will see that there are codes for handling a wide variety of tasks and calibrations. The subroutine in Figure 7.4 creates variables for each of the available codes. We will see momentarily how these variables make it easier to use rCommands.

```
InitExtendedCommands:
   MyTimeOut = 110                        // Sets timeout for 3pi
                                          // default = 500
   PololuPlay = 111                       // argument 1-4 selects
                                          // sound to play
   ServoOffset = 112                      // default is 0
   CalibrateLineSensors = 113             // argument must be provided,
                                          // but not used, sensors
                                          // automatically enabled
   SetRotationTime = 114                  // default is 12
   SetMoveTime = 115                      // default is 24
   CalibrateDrift = 116                   // default is 0 (upper signed
                                          // nibble is left wheel
                                          // lower signed nibble is
                                          // right wheel)
   CalibrateCompass = 117                 // argument must be provided,
                                          // but not used, compass is
                                          // automatically enabled
   SetTurnStyle = 118                     // default is 0 (normal
                                          // rotational turns)
                                          // numbers from 1-8 will cause
                                          // the SLOW wheel to run at
                                          // 10 times this % of the
                                          // fast wheel, example, 2
                                          // would make the slow wheel
                                          // 20% of the fast wheel.
   PololuStop = 119                       // an argument must be
                                          // provided but not used
   ResetDefaults = 120                    // an argument must be
                                          // provided but not used
   EnableCompass = 121                    // True or false, default True
                                          // If enabled, compass will be
                                          // used to augment turns
                                          // and linear travel (typical
                                          // acurracy, +/- 2 degrees
   SetHiLevel = 122                       // default is 200
   SetDistMode = 123                      //   0 - pixels (default)
                                          //   1 - inches
                                          //   2 - raw data (nonlinear)
   FindBeacon = 124                       // argument spec. direction
                                          // (1-clockwise, 0-counterCW)
   MoveToAngle = 125                      // provide 1/2 of the desire
                                          // angle (0-180)
   SetSpeed = 126                         // Sets speed of remote robot,
                                          // default 40
                                          // note: normal rSpeed
                                          //commands are ignored
   return
```

**Figure 7.4**: This subroutine creates variables whose values reflect the codes for various rCommands.

More detailed information about each of these options can be had by examining the source code in Chapter 5. Don't feel like you have to understand all of the inner workings to be able to utilize these commands in your programs. Many examples will be used in future chapters to demonstrate the power of these features. For now though, let's examine one specific example to give you a clear picture of what's to come.

Normally, a RobotBASIC program that uses the simulated robot will call a subroutine to `rLocate` the robot and perform any required initialization. We can use this routine to handle tasks such as the calibration function discussed above. The code fragments in Figure 7.5 is representative of how such a module might be implemented.

```
Main:
  CommPort=24 // set to zero to use simulation
  gosub InitRobot1
  // write your main program here
end

InitRobot1:
  gosub InitExtendedCommands    // sets up variables
  rCommPort CommPort // choose Bluetooth or simulation
  rLocate 100,100
  rCommand(CalibrateDrift, 0x10) // left wheel faster
  // place other needed rCommands here
  if CommPort=0
    // perform initialization the sim. needs
  else
    // perform initialization the real robot needs
    // example-initialize the line sensors
    print "Place the robot over a line"
    InLineInputMode ON
    input "Press ENTER when ready",a
    rCommand(CalibrateLineSensors,0)
    print "Move robot to starting location"
    input "Press ENTER when ready",a
  endif
  // place any initialization here that is for both
  // the simulated or the real robot
return
```

**Figure 7.5:** These code fragments show how one routine can initialize both the simulated robot and a real robot.

There are several things about Figure 7.5 that deserve attention. First, the use of the variable `CommPort` makes it very easy to make this program work with the simulated robot or a real robot. If `CommPort` equals zero, the simulator will be used, but if `CommPort` is set to the value of the Bluetooth port, then the external robot will be activated.

Notice that the value of `CommPort` is also tested in an `if`-statement to decide which set of initialization instructions to execute. You might prefer to place the `rCommands` in the `else`-block of this `if`-statement but it is not necessary, because RobotBASIC <u>automatically</u> ignores all `rCommands` if an external communication port has not been selected.

You might be wondering what kind of initialization might be required for only the real or the simulated robot. The simulated robot, for example, might need a special environment (like a line to follow, or objects to avoid) graphically drawn on the screen. Obviously, the real robot would not need this. On the other hand, the simulated robot does not need to have its line sensors calibrated, but the 3pi does (as discussed in earlier chapters).

The actual calibration of the line sensors is accomplished through an `rCommand` using code 113 (see Figure 7.4). Since the variable `CalibrateLineSensors` has been set to 113, we can use it in the `rCommand` of Figure 7.5 to make the code more readable. Notice also, that the user is prompted to move the robot over a line (so the robot can gather the necessary data by rotating while monitoring the line sensor values) and to press ENTER when ready. After the calibration is complete, the user is again prompted to move the robot to the desired starting position before pressing ENTER to continue with the program's operation.

Study the logic of Figure 7.5 carefully. Utilizing code structured in this way can make your programs easier to use and give them a more professional feel.

## 7.7 Dealing With Multiple Robots

There are many situations where the programs you write might have to be used with several different robots. For example, suppose you are a student in a lab or member of a robot club that has several 3pi robots. Perhaps each member of the group normally writes and debugs their program using the simulated robot. When you have your program running properly, you are given one of the real robots so that you can test your code in a real-world situation. Each of the three robots available to you are almost certainly different in many ways. For example, when moving forward, one of the robots normally might drift slightly to the left, another slightly to the right, and the third even more so to the right. In general, there would also be a variety of other calibrations that would be different for each of the real robots.

One easy solution to this problem would be to create multiple initialization modules that calibrate the real robots differently using the appropriate rCommands. If each of the robots are numbered or named, then you can simply change one gosub statement so that the appropriate initialization is performed for the robot you are currently using. In fact, a well-organized teacher or club might provide appropriate routines that are robot-specific so that each student can easily choose a routine that ensures that the robot they are using will be calibrated properly.

## 7.8 Calibration Options

Figure 7.4 reveals many calibration and initialization options available to you. For example, you can ensure that your robot turns fairly close to the requested angle, even if the compass is not used (SetRotationTime).

You can even ensure that your robot moves an appropriate forward distance (SetMoveTime). If you are using the default-sized simulated robot, a command of rForward 40, moves the robot a distance forward equal to its diameter. Ideally, the same command should move the real robot a distance equal to its diameter. Appropriate calibrations such as this, help ensure that the algorithms you develop for the simulation will work in a similar manner on the real robot.

Figure 7.4 shows numerous other options for tinkering with the robot's innate behaviors. Many of these will be examined and used in programs developed throughout the remainder of the text.

## 7.9 Summary

In this chapter you have learned:

- How to activate a real, external robot from RobotBASIC.
- Why close-loop control is better than open-loop control.
- How various actions of real robots can be calibrated to ensure they operate as desired.
- How rCommands provide versatility for customizing an external robot without having to modify its firmware.

# Chapter 8

# Line Following

Now that you know how to activate the 3pi robot from within a RobotBASIC program, it is time to make it do something interesting while demonstrating the real power of combining simulation and real-world robotics.

## 8.1 A Line-Following Algorithm

Examine the program in Figure 8.1. It produces a modestly curvy line and makes the robot follow it until the line ends. It is a very simple algorithm; one that will not work if the line bends at too sharp an angle - but certainly adequate for this example.

The width of the line in this example is important. It should be wide enough so that it can barely cover two side-by-side sensors. It will be important that the line we use for the 3pi meet the same conditions, but more on that shortly.

As stated earlier, the algorithm is very simple. We are using just three line sensors, the default mode of the simulation. The robot simply continues through a loop until it no longer sees the line (s=0). If the left sensor sees the line (s&4) then the robot turns left to get back to the line. If the right sensor sees the line (s&1) then the robot turns to the right.

```
Main:
  gosub InitLine
  gosub InitRobot
  gosub FollowLine
  print "DONE"
end

InitRobot:
  rLocate 100,300,90
  rInvisible Green
return

InitLine:
  SetColor Green
  LineWidth 5
  Line 100,300,200,300
  LineTo 300,270
  LineTo 400,300
  LineTo 500,330
  LineTo 600,400
  LineTo 700,400
return

FollowLine:
  repeat
    rForward 1
    s=rSense()
    if s&4 then rTurn -1
    if s&1 then rTurn 1
  until s=0
return

CheckSensors:
  while true
    rForward 0
    xyString 200,200,rSense()
  wend
return

// Merge Figure 7.4 here
```

**Figure 8.1:** This short program makes the simulated robot follow a relatively straight line.

## 8.2 Following a Line with the 3pi

If we are going to make the 3pi robot follow a similar line, we must first create the line. Figure 8.2 shows a line made with a black marker on white poster board. The line needs to be about 10 mm wide to have the same relative dimensions as the one used for the simulation. We will see how to test the line's width in a moment. First, let's see how to modify the program of Figure 8.1 so that it controls the real robot.

**Figure 8.2**: This is the line the real robot will follow.

In order to make the program in Figure 8.1 control the 3pi, we only need to modify the `InitRobot` subroutine. The new version of this module is shown in Figure 8.3.

```
InitRobot:
  CommPort=0
  gosub InitExtendedCommands
  rCommPort CommPort
  rLocate 100,300,90
  rInvisible Green
  if CommPort<>0
    rCommand(CalibrateLineSensors,1)
    // rCommand(SetTurnStyle,3)
    delay 4000 // give user time to position robot
  endif
return
```

**Figure 8.3**: These modifications to the original program make it control the 3pi robot instead of the simulation.

Notice the changes made in Figure 8.3. The variable `CommPort` at the beginning of the module controls everything. When that variable is zero, the program controls the simulated robot just as it did before. If you change the value though, to the port number of your Bluetooth adapter, then the program tries to control the 3pi. An important aspect to these changes is the `rCommand` that calibrates the line sensors. Remember, the 3pi's line sensors are not active until this command is executed. The calibration forces the 3pi to rotate on its axis while reading the line sensors. By recording the maximum and minimum values read and using that information to scale future data, we can be sure that the line will be recognized properly.

This means you should set the 3pi on the line (as shown in Figure 8.3) before you run the program. See Figure 7.5 more alternative prompting. When the program starts, the 3pi will rotate left and right calibrating the sensor data, and then pause for 4000 ms to give you time to move the robot to the beginning of the line (center the robot on the line).

The program should move the robot along the line and stop when the line ends. If it stops early, it may mean your line is not quite thick enough and the robot lost the line during one of its turns. The simulator will do this too. Change the width of the line to 1 and run the program in the simulation mode to see this happen. There is another reason the real robot might fail this process. See the note below for more information.

> **✓ Note:** Since the 3pi has only one caster, it can easily rock forward changing the distance the line sensors are from the line. When this happens, the data from the sensors is unreliable and may cause strange errors. It is exactly these types of errors that drive hobbyists crazy because they often assume their program itself is the problem and rewrite working code. Using the simulator first can greatly reduce the overall development time.

If you think the line thickness is the problem it can be easily checked by making a simple modification to the `Main` program. Insert the following line immediately following the `gosub InitRobot`.

```
gosub CheckSensors
```

This will force the use of a special subroutine that was shown in Figure 8.1, but not used until now. Since the new subroutine is an endless loop, the program will not try to follow the line. Instead, the endless loop in the `CheckSensors` subroutine continually displays the value of the line sensors on the screen. Note that an `rForward 0` command is also issued inside the loop. This is necessary because the sensor data is not updated unless communication is made with the 3pi. Telling the robot to move a zero distance is an easy way to force this communication.

When the program is run, manually move the robot slightly left and right. When over each of the sensors individually, the data displayed will be a 1, 2, or 4. If the line is wide enough to cover two sensors at a time, you will also see a 3 and 6 displayed as you transition between the sensors. If necessary make your real line slightly wider.

At this point, the 3pi should follow the line properly, but its movement will be slightly jerky because the robot is rotating when it turns. This rotation actually moves the robot backward when compared to the overall movement needed to follow the line. This backward movement is not noticed on the simulation, but on the real robot, the physics of inertia creates the jerky action.

## 8.3 Smoothing the Robot's Movement

We can eliminate the jerky motion by simply making the robot change the way it executes a turn. The easiest way to think about this is as follows. Normally the robot turns by moving one wheel forward while the other is reversed (causing it to pivot about its center). If we simply stopped one wheel instead of reversing it, the robot would still turn, although not quite as sharply. The important difference though, is that this new style of turn would keep the robot's motion always in the forward direction.

The 3pi's firmware supports a special rCommand that can provide a variety of turn styles. Notice the rCommand that is commented out in Figure 8.3. This causes the slow wheel (on turns) to move *forward* at 30% of the speed of the faster wheel, as described in earlier chapters. If you include this line the robot will follow the line, but with a much smoother motion.

You can actually use values from 1 to 8 to control the robot's style of turns making the slow wheel move at 10% to 80% of its normal speed. Higher numbers will give smoother operations, but the slow turn created by higher numbers might prevent the robot from being able to follow the line, especially if the line turns sharply. You can force the robot to use its normal turn style by issuing the rCommand again, but with a data parameter of 0.

It is important to realize that the 3pi firmware only uses the new turn style when the robot is asked to turn 1° forward or backward - a typical motion used during closed-loop control operations. If the 3pi is asked to turn a larger amount (90° for example) the robot will always use the default rotational style automatically.

## 8.4 Handling a Line Maze

The actual solving of a maze can be accomplished in a variety of ways. We cover some in our book, Robot Programmer's Bonanza, and the Pololu web page offers additional ideas. In most of these solutions, the robot needs to know when it has reached a turning point in the maze.

The program in Figure 8.4 shows one method for determining when the robot has encountered a left turn, a right turn and a T-crossing.

```
Main:
  gosub InitLine
  gosub InitRobot
  gosub FollowLine
  print "DONE"
end

InitRobot:
  CommPort=0
  gosub InitExtendedCommands
  rCommPort CommPort
  rLocate 100,300,90
  rSenseType 5 // enables 5 line sensors (not 3)
  rInvisible Green
  if CommPort<>0
    rCommand(CalibrateLineSensors,1)
    rCommand(SetTurnStyle,1)
    rCommand(SetMoveTime,19) // modify as needed
    delay 4000 // give user time to position robot
  endif
return

InitLine:
  SetColor Green
  LineWidth 5
  Line 100,300,200,300
  LineTo 300,270
  LineTo 400,300
  LineTo 500,330
  LineTo 600,400
  LineTo 700,400
  LineTo 690,200
  LineTo 500,210
  LineTo 510,90
return

FollowLine:
  repeat
    rForward 1
    s=rSense()
    if s&4 then rTurn -1
    if s&1 then rTurn 1
    if (s&24)=24 // T crossroad found
      // make your own choice here
      print "T found"
      rForward 0 // stop robot
      return
    elseif s&8   // right turn found
      rForward 20
      repeat
        rTurn 2
      until rSense()&2
    elseif s&16  // left turn found
      rForward 20
      repeat
        rTurn -2
      until rSense()&2
    endif
  until s=0
return

// Merge Figure 7.4 here
```

**Figure 8.4**: This program demonstrates some techniques useful for negotiating a maze.

There are several aspects of Figure 8.4 that merit further discussion. Two additions have been made to the `InitRobot` subroutine. First, the `rSenseType` command allows the robot to have 5 line sensors instead of 3. The two new sensors are to the right and left of the original 3 sensors. The new right-hand sensor is reported in bit B3 and the new left-hand sensor in B4. The original 3 sensors are still reported in bits B0-B2 as before.

Another change made to the `InitRobot` subroutine is an `rCommand` that sets the move time. You may not need this, but it is shown here for clarity. Normally, the command `rForward 20` should move the robot a distance equal to its radius - the need for this capability will be explained shortly. If your 3pi does not perform this action somewhat accurately, the Move Time should be calibrated appropriately.

Several graphical lines were added to the `InitLine` subroutine to create a series of right and left turns plus a T-crossing at the end. The goal of this program is to demonstrate how the robot can recognize these conditions and respond to them appropriately.

Normally, a line maze might be made of straight lines and right-angle turns. Since our robot uses its sensors to follow the lines, a much more general maze can be used.

Most of the modifications for this program are in the `FollowLine` subroutine copied from Figure 8.1. The actual line-following behavior has not been altered. The newly added `if-elseif-endif` statement checks the status of the 2 new line sensors to determine if a turn or T has been encountered. If a T is found, the robot is just stopped and the program terminated. The actual response would depend on how your program is solving the maze.

In the case of a left or right turn, the robot simply moves forward a distance equal to its radius, which should place it centered over the corner of the turn. It then rotates in the appropriate direction in 2° steps (if we had used 1° steps the robot would not have automatically used the rotational turn style). The robot stops its rotation when the center line sensor reaches the line. The robot then resumes following the line.

Even though the `rForward 20` command is totally open-loop control, the robot uses its sensors to find the line again, so the actual forward movement only has to approximate the proper distance. Some open-loop command can be acceptable as long as closed-loop feedback is periodically used to correct error that might have occurred.

## 8.5 Summary
In this chapter you have learned:
- ❏ How to make the 3pi robot follow a line.
- ❏ How to calibrate the line sensors on the 3pi robot.
- ❏ How to read and display sensor data from the real robot.
- ❏ How different turn styles can be invoked with an `rCommand` to make the robot move smoother.
- ❏ How two additional line sensors can be enabled to allow the robot to determine turns and T's in a line maze.

# Chapter 9

# Following a Wall

In the last chapter we learned how to make the real robot follow along a line. In this chapter we will turn that line into a wall and try to follow it just as we did the line. In this case, of course, we need to utilize the perimeter IR sensors whose data is provided by `rFeel()`.

## 9.1 Developing the Algorithm

Our first step is to use the simulator to develop an appropriate algorithm. The program in Figure 9.1 is a good start. The `InitLine` subroutine is one from Chapter 8, and the `InitRobot` should also look familiar except for the last two lines. They force the robot to leave a trail on the screen so it is easy to evaluate its performance.

The algorithm itself is all contained in the `FollowWall` subroutine. In this version, the principles are very simple. When the left most IR sensor (bit B4) is on, the robot turns away from the wall. When the next sensor to the right (B3) is off, we turn back towards the wall. This is much like a blindfolded person might follow a wall. If she moves away from the wall so that she cannot touch it she turns back toward the wall, and if she gets too close she turns away.

```
Main:
  gosub InitLine
  gosub InitRobot
  gosub FollowWall
  print "DONE"
end

InitRobot:
  CommPort=0
  gosub InitExtendedCommands
  rCommPort CommPort
  rLocate 125,325,90
  rCommand(SetTurnStyle,3)
  rInvisible green
  rPen down
return

InitLine:
  SetColor Black
  LineWidth 10
  Line 100,300,200,300
  LineTo 300,270
  LineTo 400,300
  LineTo 500,330
  LineTo 600,400
  LineTo 700,400
  LineTo 690,200
  LineTo 500,210
  LineTo 510,90
  LineWidth 2
return

FollowWall:
  while true
    rForward 1
    a=rFeel()
    if a&16 then rTurn 1
    if not(a&8) then rTurn -1
  wend
return

// Merge Figure 7.4 here
```

**Figure 9.1**: This program moves the robot along a wall.

When you run the program in Figure 9.1, you will see the output shown in Figure 9.2. The robot hugs the wall fairly well but it occasionally exhibits a loopy behavior. This is particularly a problem when rounding a sharp corner, because the robot turns back too far, causing it to head straight into the wall, which causes the error shown.

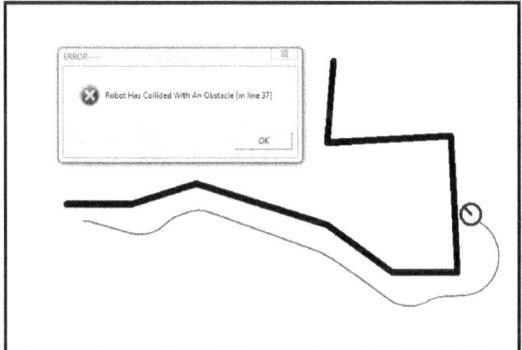

**Figure 9.2**: This is the output when the program in Figure 9.1 is executed.

We can fix this problem by having the robot check the front IR sensor and turning sharply right when it is triggered. This can be implemented by adding the following line at the end of the FollowWall subroutine (right before Wend).

```
if a&4 then rTurn 70
```

If you add the line, the new output is shown in Figure 9.3.

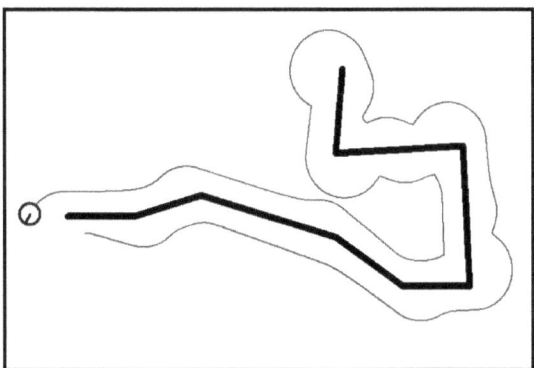

**Figure 9.3**: Now the robot no longer crashes into the wall.

As you can see from Figure 9.3, the robot now follows the wall without problems, but its path is very loopy, especially on corners and even moderately sharp turns.

After a little experimentation with the simulation (illustrating one of the advantages of having a simulator) a slightly better algorithm was developed as shown in Figure 9.4. In the new version, the robot simply makes additional turns left and right when it determines that it has drifted too far away or too close. The fact that the additional turns only happen when needed really improves the robot's overall performance as shown in Figure 9.4.

```
FollowWall:
   while true
      rForward 1
      a=rFeel()
      if a&16 then rTurn 1
      if a&8  then rTurn 1
      if not(a&8) then rTurn -1
      if not(a&24) then rTurn -1
      if a&4 then rTurn 70
   wend
return
```

**Figure 9.4**: Replace the original `FollowWall` with this routine to greatly improve the robot's performance.

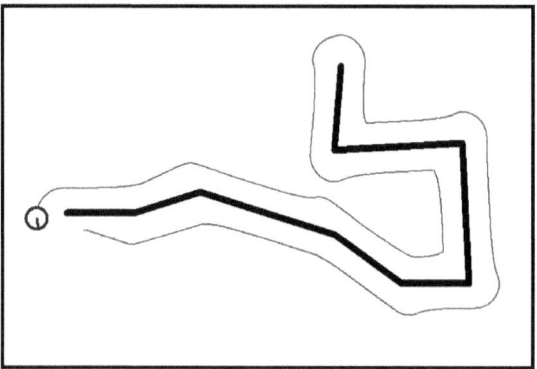

**Figure 9.5**: The robot now follows the wall perfectly.

## 9.2 Moving the Program to the 3pi

Now that the algorithm has been developed, it is easy to move this behavior to the real robot. Simply change the value of `CommPort` in the `InitRobot` subroutine to the port number of your Bluetooth adapter.

Of course, your real robot needs some kind of wall to follow as shown in Figure 9.6. Notice the walls are made from discarded Styrofoam packing. Several of these triangular shapes can form a wide variety of shapes.

**Figure 9.6**: Walls to follow can easily be made from discarded Styrofoam.

As expected, a Turn Style of 3 makes the robot hug the wall, staying about 4 inches from it. When the style was changed to 5, the robot moved noticeably smoother, but the looping (due to the slow turns) on sharp outside corners was unacceptable. An ideal solution would be for the robot to utilize slow turns on straight sections of the wall and faster turns when corners are encountered.

Figure 9.7 shows a replacement subroutine for our original program that lets the robot decide what turn style to use. A new variable `ts` keeps track of the current turn-style. Each time the robot cannot see the wall at all, `ts` is decremented. Whenever the robot is very close to the wall (either or both of the left IR sensors see the wall) then `ts` is incremented.

```
FollowWall:
  ts=3
  while true
    rForward 1
    a=rFeel()
    if a&16 then rTurn 1
    if a&8
      rTurn 1
      ts=ts+1
    endif
    if not(a&8) then rTurn -1
    if not(a&24)
      rTurn -1
      ts=ts-1
    endif
    if a&4 then rTurn 70
    if ts<3 then ts=3
    if ts>8 then ts=8
    rCommand(SetTurnStyle, ts)
  wend
return
```

**Figure 9.7**: Replace the original subroutine with this one and the robot can alter its turn-styles on its own.

These changes to the `FollowWall` subroutine allow the robot to self-adjust its turn-style as it follows the wall. The program, when run using the wall of Figure 9.6, causes the robot to appear to slow down and turn more sharply on the corners and seemingly increase its speed by using slower turns on the long straightway. As an exercise, you might consider using the `rCommand` that changes the robot's speed in a similar manner, allowing the robot to actually move faster when appropriate.

## 9.3 Using the Range Sensor

One of the great things about having a simulator is that you can try many ideas quickly to see which might have merit on a real robot. So far in this chapter, we have used the IR perimeter sensors to follow along a wall. Lets try a totally different approach.

Both our simulated robot and our 3pi have a ranging sensor on a revolving turret. Figure 9.8 shows how simple it can be to follow a wall using the ranging sensor. In this example, the robot simply turns away from the wall when distance is greater than 40 and turns back toward the wall if it is less. This means the robot continually corrects and that actually creates a very smooth behavior. If you replace the old `FollowWall` routine with this one you will see the simulated robot move as shown in Figure 9.9.

```
FollowWall:
  while true
    rForward 1
    if rRange(-90)<40
      rturn 1
    else
      rturn -1
    endif
    if (rFeel()&4) then rTurn 70
  wend
return
```

**Figure 9.8**: This new routine follows a wall using the ranging sensor.

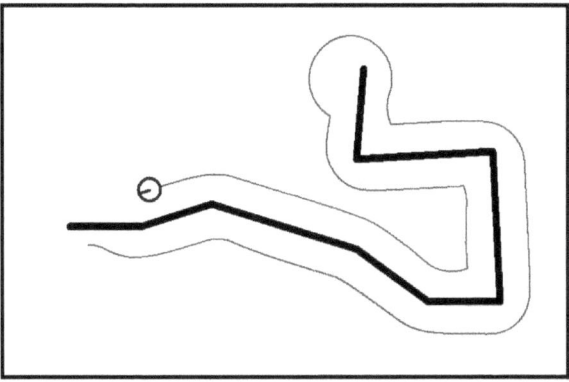

**Figure 9.9**: Wall following with the ranging sensor can be smooth.

## 9.4 Finding Doorways

Now that our robot can follow along a wall, a logical next step is for it to be able to find doorways and navigate through them. The program in Figure 9.10 makes the program follow a wall until it finds a doorway. The robot then center's itself on the opening and moves through it.

```
Main:
  gosub InitLine
  gosub InitRobot
  while 0
    print rRange()
  wend
  gosub FindDoorway
  rTurn -90    // turn
  rForward 50  // and go through door
end

InitRobot:
  SetTimeOut 30000 // compass calibration takes a long time
  CommPort=0
  gosub InitExtendedCommands
  rCommPort CommPort
  rLocate 125,325,90
  rCommand(SetTurnStyle,8)
  // use the following two commands if needed
  rCommand(CalibrateDrift,0x02)
  rCommand(CalibrateCompass,0)
  rInvisible green
  rPen down
  rSlip 15  // introduce some error for realistic development
return

InitLine:
  SetColor Black
  LineWidth 5
  Rectangle 5,5,795,520
  LineWidth 10
  line 100,300,600,300
  // now create random doorway
  RectangleWH 350+random(50),200,60+random(60),200,white,white
  LineWidth 2
return

FindDoorway:
  found=false
  while not found
    // follow the wall a short distance
    rForward 1
    a=-80
    if rRange(a)<30
      rturn 1
    else
      rturn -1
    endif
    // then check for doorway while moving
    c=0
    if rRange(a)>60
      while rRange(a)>60
        rForward 1
        c=c+1 // count the moves
      wend
      // now back up halfway
      for i=0 to (c-40)/2
        rForward -1
      next
      found=true
    endif
  wend
return

// Merge Figure 7.4 here
```

**Figure 9.10**: This program lets the robot find a
doorway and go through it.

The program in Figure 9.10 is well commented so it should be easy to understand. It draws a wall with a randomly placed and sized doorway as shown in Figure 9.11.

The algorithm for achieving our goal is deceivingly simple. Once the robot finds the doorway (by sensing a sudden increase in the distance to the wall), it simply counts how many times it moves forward before it detects the wall again. At that point, it simply backs up to the center of the doorway. The amount to back up is ½ of the distance traveled across the doorway, less the diameter of the robot, which is 40 pixels.

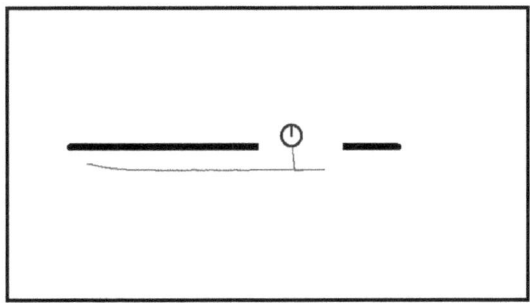

**Figure 9.11**: The robot finds the center of the doorway and moves through it, even with random error.

## 9.5 Finding the Door with the 3pi

This is a good time to discuss real-world differences and the problems they can cause. Refer to the `InitRobot` subroutine in Figure 9.10. Notice the three `rCommands` for setting the turn-style as well as calibrating drift and the compass. Let's see why these are particularly important when using this program with a real robot.

The robot needs to be moving parallel the wall when it finds the beginning of the doorway. If it is correcting at that point (turning inward, for example) then the robot will not move parallel to the wall as it crosses the doorway. The same problem occurs if the *drift* has not been corrected. This is also the reason the turn-style has been set to 8, which is a very slow turn. Because of this, the robot could not follow a curvy wall, but it follows a straight, or nearly straight wall very nicely. Both of these conditions allow the robot to move (open loop) across the short distance of the doorway without drifting to far off its parallel path.

Once the wall is seen again (on the far side of the doorway) the robot simply backs up to the middle of the doorway. The distance behind the robot (to the door frame) is the distance traveled (the value of c) less the diameter of the robot (40). We need to back up half of that distance.

As you can see from Figure 9.11, this works nicely in the simulation, even with slip added. In the real world though, there are potential problems.

First, the loop in the simulation that moves across the doorway ends up moving exactly c pixels, because the simulated robot moves forward once each time through the loop. This is almost certainly not true for our real robot. The actual distance traveled for an rForward 1 command is based on many things. Remember, in order to prevent jerky movement, the robot's motors are not turned off after an rForward 1 is executed. Instead, the robot will continue to move forward until a new command is received (or until a timeout occurs, which defaults to 500ms for our 3pi firmware).

We could try, of course, to calibrate this condition by reducing the standard speed of the 3pi so that each rForward 1 command causes it to move only 1/40 of its diameter, but most users would not be happy with the resulting sluggish behavior. Such calibration might make sense on a large robot, where 1/40 of a larger diameter might create acceptable speeds.

To that end, we need to modify the distance the robot backs up. For our 3pi, the reversing for-loop simply counted from 0 to c instead of the formula mentioned above. This worked fine over a wide range of doorway sizes.

Another point worth mentioning is the compass. If enabled, it is used to assist when turning angles larger than 20°. Since the robot turns 90° after backing up, the firmware in the 3pi will automatically record its original heading and use the compass to turn 90° to the left. This is important for two reasons.

First, if the compass is not calibrated properly (using the rCommand designed for that purpose) the angle actually turned will not be reliable. If you have a lot of electronic equipment and metal objects (tools etc.) in the area where the robot functions, it will be vital to calibrate the compass, perhaps even if you move the robot's activity to a different area of the same room. The calibrate-compass routine takes a significant amount of time, so you should generally only do it when necessary, and you must increase RobotBASIC's default timeout from 5000ms to 30000 or a timeout can occur.

If your compass is found to be unreliable for some situations, you have the option of using an rCommand to disable it. If the rotation-time has been properly calibrated (again, with an rCommand) then the robot should perform adequately. Try the program without the compass.

Secondly, the robot must back up in a straight line, reasonably parallel to the wall. If it does not, then the ending turn angle will be wrong even if the 90° is accurate. This should help emphasize the need for minimizing the robot's drift through calibration. If the 3pi had more I/O pins, we could have added wheel encoders to count the number of ticks made as the wheels revolve. Adding this option can greatly improve the open-loop movements your robot has to make. The wheel counters themselves represent a closed-loop system, but since they cannot determine if the wheels slipped on the floor (and thus did not move the robot as far as expected) we will consider them an open-loop system.

Finally, there are many hard-to-control conditions that can make a real robot respond differently from the simulation. For example, the first time the Find Door program was run with the 3pi, it failed to work properly because some readily available boxes were used for the hallway walls. The extreme thickness of walls made from boxes (and a slightly different turret angle) allowed the ranging sensor to "see" them early, making the robot think it had reached the end of the doorway prematurely. The moral: If you want your robot to find a doorway in a very thick wall, create a thick wall in your simulation so the algorithm that is developed will handle it.

No matter what you do though, there will always be the possibility that code that works perfectly on the simulator may have to be tweaked in order for it to control the real robot. That said, we think you will find it much easier to tweak code that you know *should* work, rather than spending hours trying to find errors in your logic when the real error is something simple (that is easily overlooked like a thick hallway wall).

## 9.6 Summary
In this chapter you have learned:

- ❑ How a robot can use the IR perimeter sensors to follow a wall.
- ❑ How a robot can use a ranging sensor to follow a wall.
- ❑ How a robot can locate a doorway and navigate through it.
- ❑ How various turn-styles can make a real robot's movements smoother.
- ❑ How RobotBASIC programs can alter the turn-style of the 3pi.
- ❑ How a robot can self-determine what turn-style is appropriate for the task at hand.

# Chapter 10

# Beacon Navigation

One of the things learned in Chapter 9 is that real robots do not behave perfectly. Friction, wheel slippage, and many other minor random conditions can prevent the robot from being exactly where we expect it to be. This chapter will explore the use of beacons as one potential solution for this problem.

## 10.1 Celestial Navigation

Ships at sea, especially in the past, used the stars to pinpoint their location. Sailors used a sextant to measure the angles to specific stars and then calculated the ship's position accordingly. If we place two or more beacons (to act as stars) in the corners of a room we wish to navigate, we can calculate the robot's position relative to those beacons using similar methods.

## 10.2 What is a Beacon

A beacon can be anything the robot can locate and face. The beacons detectable by our 3pi were described in Chapter 4. They can be built in many ways, the easiest of which involves a small micro controller. This task is far from complex, so nearly any processor with adequate speed will do.

Even a relatively slow controller (like a Basic Stamp, for example, can be used to create beacons as long as you have an external oscillator (perhaps even a 555). Since we had a signal generator and a Parallax BS2 micro controller handy, we used the schematic in Figure 10.1 to create multiple beacons for experimental use. You can use almost any IR diodes since the sink current of a TTL gate is minimal. A bright source of IR light is not really needed or desired. If the beacon signal is strong enough to bounce off walls, it could appear you have beacons everywhere.

The LS version of the AND gates have minimal sink currents, but they might do for simple experiments. Standard TTL gates created beacons that could be seen by the robot from eight feet or so. For longer distances add a line-driver, a buffer, or even a transistor to provide more current. You may need to add a series resistor to be sure you do not exceed the current and voltage limitations of your diodes.

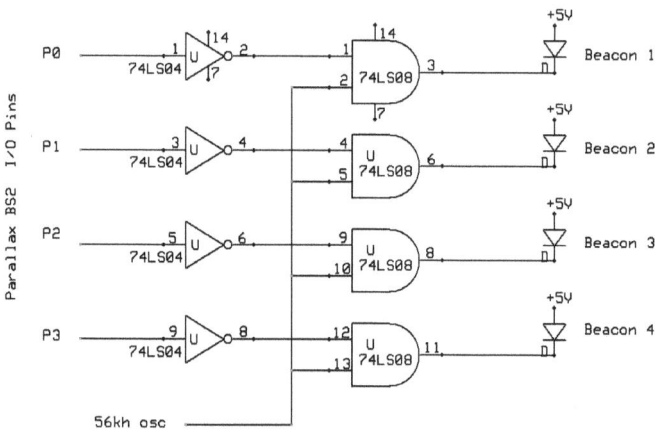

**Figure 10.1**: This simple hardware can create four experimental beacons.

Of course, you have to program the BS2 to create the proper signals on its I/O pins. The program is shown in Figure 10.2. The program is easy to create because of

Parallax's PULSOUT command that produces positive pulses in 2 microsecond intervals (PULSOUT 0,50 produces a 100 microsecond pulse on Pin 0)

```
' {$STAMP BS2}
' {$PBASIC 2.5}
' =====================================

' -----[ Pins/Constants/Variables ]-----
PulsePin          PIN       0


' -----[ Main Routine ]-----------------
DO
   PULSOUT PulsePin,50
   PULSOUT PulsePin+1,100
   PULSOUT PulsePin+2,150
   PULSOUT PulsePin+3,200
   PAUSE 10

LOOP
```

**Figure 10.2**: This BS2 program produces four beacons for experimental use.

You will have to run wires several feet or more to each IR diode, so this technique is probably not appropriate for creating individual beacons used in permanent installations, but it is perfect for experimenting with beacon applications.

## 10.3 Following a Beacon

Perhaps the easiest way to use beacons is to have them placed at strategic positions in the robot's environment. The idea is not complicated. The robot simply moves to fixed positions within the environment by finding and following the appropriate beacon signals. Let's start with a simple situation as shown in Figure 10.3

Assume we know the robot is currently in the lower left corner of the room, and that we want it to move to Position 2 in the upper right corner of the room. Also assume that the rectangle in the center of the room represents chairs, tables, and other objects that the robot would have to avoid if it tried to take the shortest path to its destination.

A less complicated solution would be to have the robot move to Position 1 first, then move to Position 2, thus allowing it to take an uncluttered path. Once we know the destinations that we want, we can hang beacons on the walls or on cabinets - any appropriate spot such that when the robot moves toward the beacon, it will pass over the desired destination.

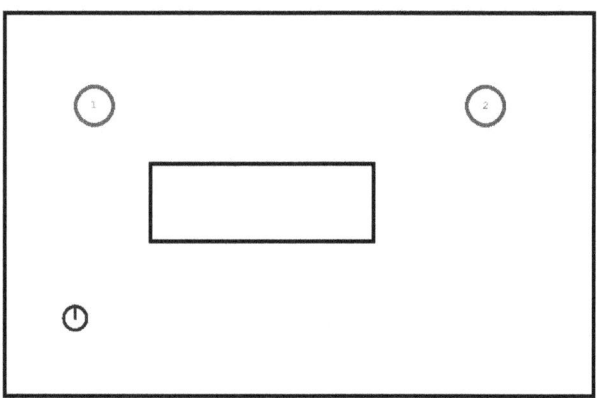

**Figure 10.3**: The robot can reach its desired destination (Position 2) by passing through Position 1.

Figure 10.4 shows how two Beacons can be positioned to allow the movements described above. Notice the beacons are not necessarily directly behind the destination circles, because the robot, when looking for the beacons may see the outer edge. This means that the turn-direction (CW or CCW) of the robot may be important (depending on the beacon detection angle).

The next step is to write a program that will make the robot look for Beacon 1, and move toward it until it gets a specified distance from the wall. At that point it stops and looks for Beacon 2. It moves toward it until it again reaches an appropriate distance from the wall. As simple as this procedure is, imagine a series of properly placed beacons throughout your home or work place. A proper data table could allow your robot to move to any of the preset destinations by taking an appropriate path from beacon to beacon. A detailed discussion of this idea can be found in our book *Robot Programmer's Bonanza* if you wish more information. For now though let's proceed with this simple example.

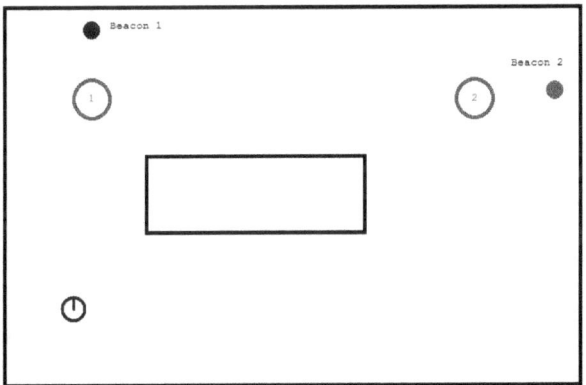

**Figure 10.4**: Properly placed beacons allow the robot to pass over the destination circles when it moves towards the beacons.

The Main Program in Figure 10.5 shows how the robot can be commanded to move to Position 1 and on to Position 2. Notice, that a new command (`call`) has been introduced in this program. Please read our HELP file whenever you are not familiar with a particular statement. You will find hundreds of commands and functions not normally associated with the BASIC dialects. The `call` statement is similar to a `gosub`, but `call` allows parameters to be passed. Also, the called modules, unlike standard subroutines, have local (instead of global) variables.

```
Main:
    gosub Init
    call MoveToBeacon(1,100)
    call MoveToBeacon(2,120)
    rForward 0 // halt
end
```

**Figure 10.5**: The program demonstrates the logic necessary to reach Position 2.

Notice how simple the logic of Figure 10.5 is to follow. After initialization, the robot is simply asked to move to Beacon 1, then on to Beacon 2. The details of how it accomplishes these tasks are dedicated to the MoveToBeacon function. Notice that two parameters are passed to MoveToBeacon. The first of these, as you probably guessed, is the number of the beacon to find and follow. The second is the distance to the wall that controls when the robot should stop its movement towards the beacon.

The details of the MoveToBeacon module is shown in Figure 10.6. The first thing you should notice is how callable modules are defined. Instead of a simple label, they start with the sub statement and have a list of the variables being passed to them in parenthesis. The logic is straightforward. First, a second routine is called that forces the robot to face the desired beacon. That routine is also included in Figure 10.6.

Once the robot is facing the beacon, it moves forward turning toward the beacon when it does not see it and away from it when it does. This closed-loop feedback allows the robot to stay on course regardless of friction and wheel slip.

```
sub MoveToBeacon(BeaconNum,DistToStop)
  call FindBeacon(BeaconNum)
  while rRange()>DistToStop
    rForward 1
    if rBeacon(BeaconNum)
      rTurn 1
    else
      rTurn -1
    endif
  wend
return

Sub FindBeacon(b)
  while not rBeacon(b)
    rturn -1
  wend
return
```

**Figure 10.6**: Following the beacon is easier than you might imagine.

Of course, you still need the initialization module. It is shown in Figure 10.7. Also, you will need to merge Figure 7.4.

```
Init:
    gosub InitExtendedCommands
    CommPort = 0
    rCommPort CommPort
    // create beacons
    CircleWH 115,25,20,20,1,1
    xyString 150,20,"Beacon 1"
    CircleWH 750,105,20,20,2,2
    xyString 700,70,"Beacon 2"
    // destination circles
    Circle 100,100,150,150,Red
    Circle 625,100,675,150,Red
    // simulate clutter in room
    rectangle 200,200,500,300,Black
    // setup robot
    rLocate 100,400
    rInvisible 1,2,Red
    rCommand(SetTurnStyle,8)
return
```

**Figure 10.7**: This routine prepares the environment.

If you run the assembled program, you will see that it always ends up in the desired destination. Now it is time to move the algorithm to the real robot.

## 10.4 Following a Beacon with the 3pi

Only a couple routines need to be modified to allow the algorithm to work well with the 3pi robot. In the Main program of Figure 10.5, the two distances were both changed to 50 due to the small work area used for our real-robot environment. The new FindBeacon routine is shown in Figure 10.8.

The original algorithm for finding the beacon (Figure 10.6) will work with the 3pi but only in a very sluggish manner. The robot must move very slowly to find the beacon because the Bluetooth interface prevents the RobotBASIC program from quickly stopping the robot when a beacon is detected. To remedy this problem we added a special rCommand that allows the robot to look for the beacon on its own. Since there is no communication delay, the robot can turn quickly and stop immediately when a beacon is detected.

```
Sub FindBeacon(b)
  if _CommPort<>0
    while not rBeacon(b)
      rCommand(_FindBeacon,0)
    wend
  else
    while not rBeacon(b)
      rturn -1
    wend
  endif
return
```

**Figure 10.8**: A special rCommand improves the robot's performance when finding a beacon.

The rCommand for finding the beacon moves to the next beacon detected. This means that the user should then use rBeacon to determine if the beacon found is the one desired. There are several aspects of this rCommand that deserve discussion. First, the argument passed determines the direction that the robot will rotate to find the beacon (0-CCW, 1-CW). Also, the first byte in the five bytes returned (all rCommands return 5 bytes) specifies which beacon was actually found, giving you two ways to determine that.

Finally, it is import to know that this rCommand always moves away from any beacon that it is pointing at, before starting to look for another. This makes it easy to move through several beacons looking for a specific one.

Figure 10.8 shows how the rCommand for finding a beacon is used to improve the performance of the 3pi during the execution of this algorithm.

Both beacons for the real robot were all created from IR LED's (see Figure 10.9) using a BS2 as described earlier. The AND gates simply enabled the 56kh signal from the generator shown in Figure 10.10.

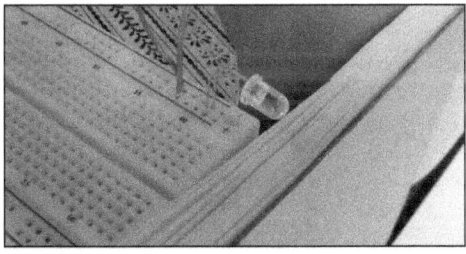

**Figure 10.9**: A beacon is a properly pulsing IR LED.

**Figure 10.10**: A Generator produces the 56kh signal
that is gated by the BS2 program.

Figure 10.11 shows the overall layout of the robot's environment. The book in the center serves as the room's clutter. The robot is expected to travel from right to left as pictured, moving toward the first beacon. When it reaches the "wall", it turns and moves toward the second beacon.

A successful move should place the robot in the destination circle shown in the figure. Figure 10.12 shows the accuracy achieved from a typical move (note the robot's position compared to the desired destination circle shown in Figure 10.11). It is slightly off, but remember, while the robot may be off slight at each destination, the error does not accumulate over time.

**Figure 10.11**: This work area lets the robot move similar to the simulated environment of Figure 10.4.

**Figure 10.12**: This represents the accuracy typically demonstrated by the robot.

Our simple two-beacon environment demonstrates the principles needed to allow a robot to easily move around your home or office.

## 10.5 A Local Positioning System (LPS)

GPS systems (Global Position Systems) like those used in your car, are valuable tools and can certainly be used on mobile robots, although the accuracy is often limited to 20 feet or more. Couple that with the fact that GPS signals are often difficult to receive inside buildings and it becomes apparent we need additional choices.

Our robot's beacon detection capabilities provide us with the basis for implementing an LPS (Local Positioning System). Turn your attention to Figure 10.13.

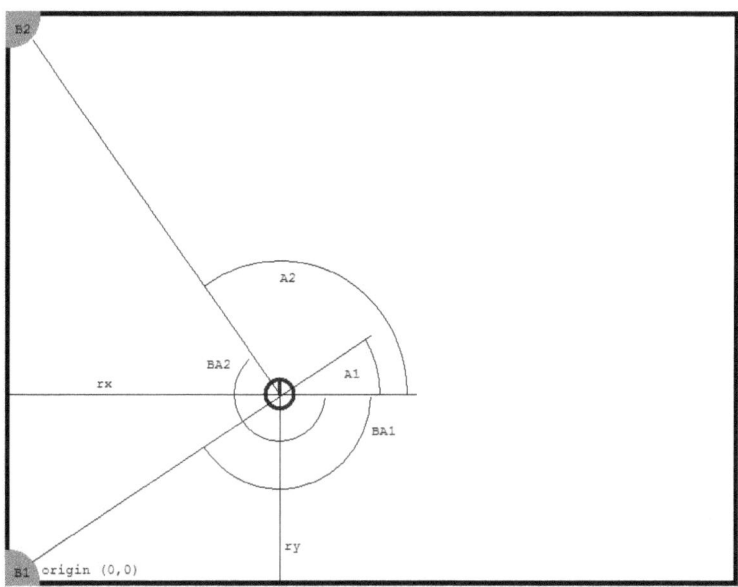

**Figure 10.13**: These angles allow us to analyze the mathematics of a simple LPS.

Figure 10.13 shows a room with two beacons, labeled **B1** and **B2**. The robot's position in the room, relative to the origin is rx, ry. If we have the robot locate and face Beacon **B1**, its current compass reading will be angle BA1. Likewise, the angle measured to **B2** would be BA2.

We can use trigonometry to convert these angular measurements to the robot's position in the room ($rx$, $ry$). The math is a little easier to see, if we use the angles $A1$ and $A2$ for our calculations. We can transform the angles as shown below.

```
A1 = 180-BA1
A2 = 360-BA2
```

The slope of each of the lines (we will call them $M1$ and $M2$) can be calculated as follows:

```
M1 = tan(A1)
M2 = tan(A2)
```

The equation for a line has the general form of

```
y=mx+b
```

where **m** is the slope of the line and **b** is the position where the line intercepts the y-axis.

The equation for both lines (Line 1 is associated with Beacon 1 and Line 2 is associated with Beacon 2) are shown below. The variable $RW$ is the room width, or the distance between the two beacons.

```
y = M1*x + 0
y = M2*x + RW
```

Since both equations are equal to y, we can say:

```
M1*x = M2*x + RW
```

Solving for x, we get:

```
x = RW/(M1-M2)
```

Once we have the value for x, it is easy to obtain y using one of the previous line equations.

```
y = M1*x
```

And as complicated as it sounds, these last two equations are all that is necessary to calculate the **x,y** position of our robot in a room. And all that we need is the two angles

(BA1 and BA2) to the beacons from the robot's current position.

Let's see how easy it is to implement these equations with RobotBASIC. The MainProgram, and the Initialization subroutine are shown in Figure 10.14.

```
MainProgram:
   gosub Initialization
   gosub FindAngles
   gosub FindXY
end

Initialization:
   CommPort=24
   gosub InitExtendedCommands
   rCommPort CommPort
   rLocate 10,10
   RA=38 // room offset angle
   RoomWidth = 52
return
```

**Figure 10.14**: The MainProgram calls other subroutines to accomplish its job.

The Initialization Module creates two important variables. The RoomWidth variable we have already discussed. It should be set to the distance between the two beacons. In my experimental environment the distance was 52 inches. In general, the further the beacons are apart, the more accurate the LPS will be. The second important variable is RA, the *room angle*. It is unlikely that your room walls just happen to be aligned with due north. For that reason, you must use an offset variable that can translate the compass readings into angles that correspond with the room's orientation. The value of this variable must be accurate. Even a few degrees of error here can cause major errors in the final calculations.

Figure 10.15 shows the FindAngles subroutine. In the body of this routine is a while-loop whose purpose is to find each of the beacons. It continues until both have been found and measured. The two variables that hold the angles for Beacons 1 and 2 are BA1 and BA2, and are both initially set to -1, to indicate that no angle has been established.

The `while`-loop will continue to execute as long as either of these variables is still -1. Inside the loop, an `rCommand` is used to locate the next beacon. Two `if`-statements check to see which beacon was found, and set the angle variables for that beacon. Note that the room offset is subtracted here and the angles adjusted if they happen to become negative.

```
FindAngles:
  BA1 = -1
  BA2 = -1
  while (BA1=-1) or (BA2=-1)
    rCommand(FindBeacon,0)
    if rBeacon(1)
      BA1=rCompass()-RA
      if BA1<0 then BA1=BA1+360
    endif
    if rBeacon(3)
      BA2=rCompass()-RA
      if BA2<0 then BA2=BA2+360
    endif
  wend
  A1=180-BA1
  A2=360-BA2
return
```

**Figure 10.15**: This routine allows the robot to find both beacons and record the angles associated with each.

At the end of the routine, the measured angles are converted to `A1` and `A2` just as we did in the mathematical analysis above. Now that we have the angles, it is easy to find the robot's position in the room.

Figure 10.16 shows the `FindXY` subroutine. It follows the mathematics discussed earlier. The routine `DtoR()` is used to convert degrees (used by the robot) to radians (required for the `tan()` function).

```
FindXY:
  M1=tan(DtoR(A1))
  M2=tan(DtoR(A2))
  rx=RoomWidth/(M1-M2)
  ry=M1*rx
  print rx;ry
return
```

**Figure 10.16**: Once the angles have been found, calculating the robot's position is easy.

## 10.6 Real-World Limitations.

As mentioned earlier, the room angle must be established accurately as it *greatly* influences the accuracy of this routine.

Also, if the room is fairly large, you will need a diode cluster for each beacon (instead of a single diode) so that the beacon can be detected from all angles.

When the angles to the beacons get very small or very large, very tiny changes in the angle will result in dramatic changes in the calculations. For that reason, you should not expect a high degree of accuracy around the perimeter of the room; that is near the walls. In the interior of the room though, our 3pi could often locate itself within six inches and nearly always within a foot. While this may not seem impressive, remember that a standard GPS is often only accurate to twenty feet or more.

If you need better accuracy, you might place 3 beacons in a room and calculate the position using two of the beacons at a time. The different calculations can then be averaged to get a final answer.

Once your application program determines approximately where the robot is, it can use the ranging sensor to measure the robot's location compared to walls, and so forth, allowing your robot to further determine its position in the room.

## 10.7 The Beacon Detector

The construction of the beacon detector housing was discussed in Chapter 2, but note the following considerations. The slit that allows light to reach the photo-sensor should be thin and made of dark material. We used black foam to line the inside of the housing and left an opening about 1/8 inch thick. If it is too thin, the robot might overshoot the beam when it looks for a beacon, especially when the robot is far from the beacon.

The slit should be fairly wide from top to bottom, allowing the robot to see beacons at different heights. The slit should also be an inch or so deep so that light cannot enter the cavity from other angles.

If the slit is too wide, then the robot will see the beacon before it actually turns *directly* at it. This means that two different angles can be obtained depending on whether the robot approaches the beacon from the left or the right.

The rCommand that looks for the beacons in the FindAngles subroutine turns the robot CCW. Change the parameter from 0 to 1 and the robot will search for the beacons using a clockwise turn.

Our slit was narrow enough and deep enough that our robot gave similar values when the robot turned CCW or CW. If your slit is too wide, then the two sets of reading will be different and should be averaged together to obtain a final answer.

## 10.8 Summary

In this chapter you have learned:

- ❑ How beacons can be used to find the position of a robot in a room.
- ❑ Why mathematics is important in fields such as robotics.
- ❑ How trigonometry can be used to create a navigation system.
- ❑ What subjects you should study if you want a career in robotics.

# Chapter 11

# What's Next?

O ur 3pi journey has finally come to an end. We hope you have enjoyed it as much as we have, because new endeavors can now begin.

## 11.1 A RobotBASIC Robot

The 3pi project showed conclusively that a robot can be built with nearly all of the capabilities of the RobotBASIC simulated robot. Furthermore, it showed that the very same programs developed to control the simulated robot can often be used to control the real robot with no modification and even when changes are required they are minimal.

Once the 3pi has been modified and programmed as outlined in Chapters 2 to 6, there should be no need to program it again at the microcontroller level. This greatly opens the field of hobby robotics to many more people. It also makes it easy for seasoned hobbyists to turn their attention to programming intelligent behaviors for their robot instead of spending their time interfacing and troubleshooting hardware and firmware.

We recognize that the construction of the 3pi as outlined in this book, requires a significant amount of skill. Ideally, if there is enough demand, Pololu will consider offering kits to make the process easier.

## 11.2 A RobotBASIC ROS

Creating a robot capable of performing nontrivial tasks requires programming skills. In the past, most hobbyists have been so bogged down building a robot that they seldom were able to give programming the attention it deserves. This is not so different from the early 1970's when hobbyists had to build their own PC's. After Apple, Radio Shack, and others introduced fully assembled computers, users were finally able to become programmers, creating applications that have changed the world.

We at RobotBASIC feel such a landmark change is about to occur in the field of robotics and we are excited that RobotBASIC will be a major impetus towards this change.

Our goal is to produce a RobotBASIC compatible, Robot Operating System. This ROS would allow motors and sensors from many vendors to be connected together to quickly and easily assemble the electronics needed for a RobotBASIC compatible robot. The robot could be small and portable or a man-sized machine capable of addressing real-world tasks.

The hardware to build sophisticated robots is readily available for those with a background in electronics. If robotic hardware becomes plug and play, allowing robots to be built by everyone, we believe hobbyists everywhere will be able to create and share RobotBASIC libraries that will allow both the art and science of robotics to flourish like never before.

Watch our web page. We will keep you informed of our progress.

www.RobotBASIC.com

# Index

3pi,

www.ingramcontent.com/pod-product-compliance
Lightning Source LLC
Chambersburg PA
CBHW051527170526
45165CB00002B/642